HUMAN AUGMENTATION AND HUMAN-MACHINE COLLABORATION

HUMAN AUGMENTATION AND HUMAN-MACHINE COLLABORATION

Tomislav Pribanic

Kruger Brentt
Publishers
2025

Kruger Brentt Publishers UK. LTD.
Company Number 9728962

Regd. Office: 68 St Margarets Road, Edgware, Middlesex HA8 9UU

ISBN: 978-1-78715-403-2

Disclaimer:

Every possible effort has been made to ensure that the information contained in this book is accurate at the time of going to press, and the publisher and author cannot accept responsibility for any errors or omissions, however caused. No responsibility for loss or damage occasioned to any person acting, or refraining from action, as a result of the material in this publication can be accepted by the editor, the publisher or the author. The Publisher is not associated with any product or vendor mentioned in the book. The contents of this work are intended to further general scientific research, understanding and discussion only. Readers should consult with a specialist where appropriate.

Every effort has been made to trace the owners of copyright material used in this book, if any. The author and the publisher will be grateful for any omission brought to their notice for acknowledgement in the future editions of the book.

For information on all our publications visit our website at http://krugerbrentt.com/

PREFACE

Human augmentation and human-machine collaboration represent transformative fields at the intersection of technology and human capability enhancement. These areas explore how advancements in robotics, artificial intelligence (AI), neurotechnology, and biotechnology can augment human physical, cognitive, and sensory abilities, as well as facilitate synergistic interactions between humans and machines to achieve tasks more efficiently and effectively than either could alone.

Human augmentation focuses on enhancing human abilities beyond natural biological limits through technological interventions. Advanced prosthetic limbs and wearable exoskeletons restore mobility and strength to individuals with limb loss or physical disabilities. These technologies incorporate sensors, actuators, and AI algorithms to mimic natural movements and provide greater functionality and independence. Neuroprosthetics directly interface with the nervous system, enabling individuals with spinal cord injuries or neurological disorders to control devices such as robotic limbs using neural signals. BCIs allow for direct communication between the brain and external devices, enhancing communication, mobility, and quality of life. Cognitive augmentation technologies, such as AI-driven cognitive prosthetics and neurostimulation techniques, aim to enhance memory, attention, and decision-making abilities. These innovations hold potential in fields like education, healthcare, and professional training.

Human-machine collaboration focuses on harnessing the complementary strengths of humans and machines to achieve synergistic outcomes. Industrial robots automate repetitive tasks in manufacturing and assembly, improving efficiency, precision, and workplace safety. Collaborative robots (cobots) work alongside humans, enhancing productivity and flexibility in shared workspaces. AI algorithms analyze vast datasets, optimize processes, and make predictions with speed and accuracy. In fields like healthcare, finance, and logistics, AI-driven insights support decision-making, diagnostics, and resource management. AR overlays digital information onto the physical world, enhancing real-time decision-making and training simulations. VR immerses users in interactive environments, facilitating experiential learning and skill development in diverse domains.

The integration of human augmentation and human-machine collaboration poses ethical, societal, and regulatory challenges. Ethical considerations include privacy concerns, equitable access to technologies, and the impact on human identity and autonomy. Addressing these issues requires thoughtful policy frameworks and stakeholder engagement. These technologies offer potential benefits such as improved healthcare outcomes, enhanced workplace safety, and increased accessibility for individuals with disabilities. They also foster innovation, economic growth, and global competitiveness. As these technologies evolve, there is a growing demand for education and training to equip individuals with the skills necessary to work alongside advanced technologies effectively. Lifelong learning and upskilling initiatives are crucial for preparing the workforce for the future.

Tomislav Pribanic

CONTENTS

CHAPTER 1

INTRODUCTION TO HUMAN AUGMENTATION AND COLLABORATION WITH MACHINE

1.1 THE EVOLUTION OF HUMAN-MACHINE INTERACTION

Human-machine interaction (HMI) has undergone a profound evolution since its inception, transforming from rudimentary interfaces to sophisticated, intuitive systems. This journey reflects advances in technology, changes in user expectations, and the growing integration of machines into everyday life.

The evolution of human-machine interaction has been marked by a series of transformative milestones, each building on the advancements of its predecessors. From the mechanical interfaces of the 19th century to the AI-driven technologies of today, HMI has continually evolved to become more intuitive, natural, and integrated into our daily lives. As we look to the future, the continued development of AI, voice assistants, gesture controls, and brain-computer interfaces promises to further revolutionize how we interact with machines, offering unprecedented levels of personalization and immersion. However, with these advancements come new challenges and responsibilities, highlighting the need for thoughtful consideration of the ethical and societal impacts of emerging HMI technologies.

1.1.1 Early Mechanical Interfaces

The Birth of HMI

The history of human-machine interaction dates back to the early 19th century, with the invention of mechanical devices like the Jacquard loom. This loom used punch cards to control the weaving of patterns in textiles, representing one of the first instances of programmable machinery. These early machines required significant manual intervention, with interfaces that were cumbersome and difficult to use.

The Advent of Typewriters and Telegraphs

The late 19th and early 20th centuries saw the introduction of typewriters and telegraphs, which significantly enhanced human-machine interaction. Typewriters standardized the input method for text, making it easier for humans to communicate with machines. The telegraph, on the other hand, allowed for long-distance communication, though it relied on operators proficient in Morse code. These devices laid the groundwork for more sophisticated interfaces by establishing fundamental interaction paradigms.

1.1.2 The Era of Mainframes and Command-Line Interfaces

Mainframe Computers

A mainframe computer, informally called a mainframe or big iron, is a computer used primarily by large organizations for critical applications like bulk data processing for tasks such as censuses, industry and consumer statistics, enterprise resource planning, and large-scale transaction processing. A mainframe computer is large but not as large as a supercomputer and has more processing power than some other classes of computers, such as minicomputers, servers, workstations, and personal computers. Most large-scale computer-system architectures were established in the 1960s, but they continue to evolve. Mainframe computers are often used as servers.

The term *mainframe* was derived from the large cabinet, called a *main frame*, that housed the central processing unit and main memory of early computers. Later, the term *mainframe* was used to distinguish high-end commercial computers from less powerful machines.

Design

Modern mainframe design is characterized less by raw computational speed and more by:

- Redundant internal engineering resulting in high reliability and security
- Extensive input-output ("I/O") facilities with the ability to offload to separate engines
- Strict backward compatibility with older software
- High hardware and computational utilization rates through virtualization to support massive throughput
- Hot swapping of hardware, such as processors and memory

The high stability and reliability of mainframes enable these machines to run uninterrupted for very long periods of time, with mean time between failures (MTBF) measured in decades.

Mainframes have high availability, one of the primary reasons for their longevity, since they are typically used in applications where downtime would be costly or catastrophic. The term reliability, availability and serviceability (RAS) is a defining characteristic of mainframe computers. Proper planning and implementation are required to realize these features. In addition, mainframes are more secure than other computer types: the NIST vulnerabilities database, US-CERT, rates traditional mainframes such as IBM Z (previously called z Systems, System z, and zSeries), Unisys Dorado, and Unisys Libra as among the most secure, with vulnerabilities in the low single digits, as compared to thousands for Windows, UNIX, and Linux. Software upgrades usually require setting up the operating system or portions thereof, and are non disruptive only when using virtualizing facilities such as IBM z/OS and Parallel Sysplex, or Unisys XPCL, which support workload sharing so that one system can take over another's application while it is being refreshed.

In the late 1950s, mainframes had only a rudimentary interactive interface (the console) and used sets of punched cards, paper tape, or magnetic tape to transfer data and programs. They operated in batch mode to support back office functions such as payroll and customer billing, most of which were based on repeated tape-based sorting and merging operations followed by line printing to preprinted continuous stationery.

When interactive user terminals were introduced, they were used almost exclusively for applications (e.g., airline booking) rather than program development. However, in 1961 the first academic, general-purpose timesharing system that supported software development, CTSS, was released at MIT on an IBM 709, later 7090 and 7094. Typewriter and Teletype devices were common control consoles for system operators through the early 1970s, although ultimately supplanted by keyboard/display devices.

By the early 1970s, many mainframes acquired interactive user terminals operating as timesharing computers, supporting hundreds of users simultaneously along with batch processing. Users gained access through keyboard/typewriter terminals and later character-mode text terminal CRT displays with integral keyboards, or finally from personal computers equipped with terminal emulation software.

By the 1980s, many mainframes supported general purpose graphic display terminals, and terminal emulation, but not graphical user interfaces. This form of end-user computing became obsolete in the 1990s due to the advent of personal computers provided with GUIs.

After 2000, modern mainframes partially or entirely phased out classic "green screen" and color display terminal access for end-users in favour of Web-style user interfaces.

The infrastructure requirements were drastically reduced during the mid-1990s, when CMOS mainframe designs replaced the older bipolar technology. IBM claimed that its newer mainframes reduced data center energy costs for power and cooling, and reduced physical space requirements compared to server farms.

Characteristics

Modern mainframes can run multiple different instances of operating systems at the same time. This technique of virtual machines allows applications to run as if they were on physically distinct computers. In this role, a single mainframe can replace higher-functioning hardware services available to conventional servers. While mainframes pioneered this capability, virtualization is now available on most families of computer systems, though not always to the same degree or level of sophistication.

Mainframes can add or hot swap system capacity without disrupting system function, with specificity and granularity to a level of sophistication not usually available with most server solutions. Modern mainframes, notably the IBM Z servers, offer two levels of virtualization: logical partitions (LPARs, via the PR/SM facility) and virtual machines (via the z/VM operating system). Many mainframe customers run two machines: one in their primary data center and one in their backup data center—fully active, partially active, or on standby—in case there is a catastrophe affecting the first building. Test, development, training, and production workload for applications and databases can run on a single machine, except for extremely large demands where the capacity of one machine might be limiting. Such a two-mainframe installation can support continuous business service, avoiding both planned and unplanned outages. In practice, many customers use multiple mainframes linked either by Parallel Sysplex and shared DASD (in IBM's case), or with shared, geographically dispersed storage provided by EMC or Hitachi.

Mainframes are designed to handle very high volume input and output (I/O) and emphasize throughput computing. Since the late 1950s, mainframe designs have included subsidiary hardware (called *channels* or *peripheral processors*) which manage the I/O devices, leaving the CPU free to deal only with high-speed memory. It is common in mainframe shops to deal with massive databases and files. Gigabyte to terabyte-size record files are not unusual. Compared to a typical PC, mainframes commonly have hundreds to thousands of times as much data storage online, and can access it reasonably quickly. Other server families also offload I/O processing and emphasize throughput computing.

Mainframe return on investment (ROI), like any other computing platform, is dependent on its ability to scale, support mixed workloads, reduce labor costs, deliver uninterrupted service for critical business applications, and several other risk-adjusted cost factors.

Mainframes also have execution integrity characteristics for fault tolerant computing. For example, z900, z990, System z9, and System z10 servers effectively execute result-oriented instructions twice, compare results, arbitrate between any differences (through instruction retry and failure isolation), then shift workloads "in flight" to functioning processors, including spares, without any impact to operating systems, applications, or users. This hardware-level feature, also found in HP's NonStop systems, is known as lock-stepping, because both processors take their "steps" (i.e., instructions) together. Not all applications absolutely need the assured integrity that these systems provide, but many do, such as financial transaction processing.

Current Market

IBM, with the IBM Z series, continues to be a major manufacturer in the mainframe market. In 2000, Hitachi co-developed the zSeries z900 with IBM to share expenses, and the latest Hitachi AP10000 models are made by IBM. Unisys manufactures ClearPath Libra mainframes, based on earlier Burroughs MCP products and ClearPath Dorado mainframes based on Sperry Univac OS 1100 product lines. Hewlett Packard Enterprise sells its unique NonStop systems, which it acquired with Tandem Computers and which some analysts classify as mainframes. Groupe Bull's GCOS, Stratus OpenVOS, Fujitsu (formerly Siemens) BS2000, and Fujitsu-ICL VME mainframes are still available in Europe, and Fujitsu (formerly Amdahl) GS21 mainframes globally. NEC with ACOS and Hitachi with AP10000-VOS3 still maintain mainframe businesses in the Japanese market.

The amount of vendor investment in mainframe development varies with market share. Fujitsu and Hitachi both continue to use custom S/390-compatible processors, as well as other CPUs (including POWER and Xeon) for lower-end systems. Bull uses a mixture of Itanium and Xeon processors. NEC uses Xeon processors for its low-end ACOS-2 line, but develops the custom NOAH-6 processor for its high-end ACOS-4 series. IBM also develops custom processors in-house, such as the Telum. Unisys produces code compatible mainframe systems that range from laptops to cabinet-sized mainframes that use homegrown CPUs as well as Xeon processors. Furthermore, there exists a market for software applications to manage the performance of mainframe implementations. In addition to IBM, significant market competitors include BMC and Precisely,; former competitors include Compuware and CA Technologies. Starting in the 2010s, cloud computing is now a less expensive, more scalable alternative.

The Backdrop

Several manufacturers and their successors produced mainframe computers from the 1950s until the early 21st century, with gradually decreasing numbers and a gradual transition to simulation on Intel chips rather than proprietary hardware. The US group of manufacturers was first known as "IBM and the Seven Dwarfs": usually Burroughs, UNIVAC, NCR, Control Data, Honeywell, General Electric and RCA, although some lists varied. Later, with the departure of General Electric and

RCA, it was referred to as IBM and the BUNCH. IBM's dominance grew out of their 700/7000 series and, later, the development of the 360 series mainframes. The latter architecture has continued to evolve into their current zSeries mainframes which, along with the then Burroughs and Sperry (now Unisys) MCP-based and OS1100 mainframes, are among the few mainframe architectures still extant that can trace their roots to this early period. While IBM's zSeries can still run 24-bit System/360 code, the 64-bit IBM Z CMOS servers have nothing physically in common with the older systems. Notable manufacturers outside the US were Siemens and Telefunken in Germany, ICL in the United Kingdom, Olivetti in Italy, and Fujitsu, Hitachi, Oki, and NEC in Japan. The Soviet Union and Warsaw Pact countries manufactured close copies of IBM mainframes during the Cold War; the BESM series and Strela are examples of independently designed Soviet computers. Elwro in Poland was another Eastern Bloc manufacturer, producing the ODRA, R-32 and R-34 mainframes.

Shrinking demand and tough competition started a shakeout in the market in the early 1970s—RCA sold out to UNIVAC and GE sold its business to Honeywell; between 1986 and 1990 Honeywell was bought out by Bull; UNIVAC became a division of Sperry, which later merged with Burroughs to form Unisys Corporation in 1986.

In 1984 estimated sales of desktop computers ($11.6 billion) exceeded mainframe computers ($11.4 billion) for the first time. IBM received the vast majority of mainframe revenue. During the 1980s, minicomputer-based systems grew more sophisticated and were able to displace the lower end of the mainframes. These computers, sometimes called *departmental computers*, were typified by the Digital Equipment Corporation VAX series.

In 1991, AT&T Corporation briefly owned NCR. During the same period, companies found that servers based on microcomputer designs could be deployed at a fraction of the acquisition price and offer local users much greater control over their own systems given the IT policies and practices at that time. Terminals used for interacting with mainframe systems were gradually replaced by personal computers. Consequently, demand plummeted and new mainframe installations were restricted mainly to financial services and government. In the early 1990s, there was a rough consensus among industry analysts that the mainframe was a dying market as mainframe platforms were increasingly replaced by personal computer networks. InfoWorld's Stewart Alsop infamously predicted that the last mainframe would be unplugged in 1996; in 1993, he cited Cheryl Currid, a computer industry analyst as saying that the last mainframe "will stop working on December 31, 1999", a reference to the anticipated Year 2000 problem (Y2K).

That trend started to turn around in the late 1990s as corporations found new uses for their existing mainframes and as the price of data networking collapsed in most parts of the world, encouraging trends toward more centralized computing. The growth of e-business also dramatically increased the number of back-end transactions processed by mainframe software as well as the size and throughput of databases. Batch processing, such as billing, became even more important (and larger) with the growth of e-business, and mainframes are particularly adept at large-scale batch computing.

Another factor currently increasing mainframe use is the development of the Linux operating system, which arrived on IBM mainframe systems in 1999. Linux allows users to take advantage of open source software combined with mainframe hardware RAS. Rapid expansion and development in emerging markets, particularly People's Republic of China, is also spurring major mainframe investments to solve exceptionally difficult computing problems, e.g., providing unified, extremely high volume online transaction processing databases for 1 billion consumers across multiple industries (banking, insurance, credit reporting, government services, etc.).

In late 2000, IBM introduced 64-bit z/Architecture, acquired numerous software companies such as Cognos and introduced those software products to the mainframe. IBM's quarterly and annual reports in the 2000s usually reported increasing mainframe revenues and capacity shipments. However, IBM's mainframe hardware business has not been immune to the recent overall downturn in the server hardware market or to model cycle effects.

For example, in the 4th quarter of 2009, IBM's System z hardware revenues decreased by 27% year over year. But MIPS (millions of instructions per second) shipments increased 4% per year over the past two years. Alsop had himself photographed in 2000, symbolically eating his own words ("death to the mainframe").

In 2012, NASA powered down its last mainframe, an IBM System z9. However, IBM's successor to the z9, the z10, led a New York Times reporter to state four years earlier that "mainframe technology—hardware, software and services—remains a large and lucrative business for I.B.M., and mainframes are still the back-office engines behind the world's financial markets and much of global commerce". As of 2010, while mainframe technology represented less than 3% of IBM's revenues, it "continue[d] to play an outsized role in Big Blue's results".

IBM has continued to launch new generations of mainframes: the IBM z13 in 2015, the z14 in 2017, the z15 in 2019 and the z16 in 2022, the latter featuring among other things an "integrated on-chip AI accelerator" and the new Telum microprocessor.

Differences from Supercomputers

A supercomputer is a computer at the leading edge of data processing capability, with respect to calculation speed. Supercomputers are used for scientific and engineering problems (high-performance computing) which crunch numbers and data, while mainframes focus on transaction processing. The differences are:

- Mainframes are built to be reliable for transaction processing (measured by TPC-metrics; not used or helpful for most supercomputing applications) as it is commonly understood in the business world: the commercial exchange of goods, services, or money. A typical transaction, as defined by the Transaction Processing Performance Council, updates a database system for inventory control (goods), airline reservations (services), or banking (money) by adding a record. A transaction may refer to a set of operations including disk read/writes, operating system calls, or some form of data transfer from one subsystem to another which is not measured by the processing speed of the CPU. Transaction processing is not exclusive to mainframes but is also used by microprocessor-based servers and online networks.
- Supercomputer performance is measured in floating point operations per second (FLOPS) or in traversed edges per second or TEPS, metrics that are not very meaningful for mainframe applications, while mainframes are sometimes measured in millions of instructions per second (MIPS), although the definition depends on the instruction mix measured. Examples of integer operations measured by MIPS include adding numbers together, checking values or moving data around in memory (while moving information to and from storage, so-called I/O is most helpful for mainframes; and within memory, only helping indirectly). Floating point operations are mostly addition, subtraction, and multiplication (of *binary* floating point in supercomputers; measured by FLOPS) with enough digits of precision to model continuous phenomena such as weather prediction and nuclear simulations (only recently standardized *decimal* floating point, not used in supercomputers, are appropriate for monetary values such as those useful for mainframe applications). In terms of computational speed, supercomputers are more powerful.

Mainframes and supercomputers cannot always be clearly distinguished; up until the early 1990s, many supercomputers were based on a mainframe architecture with supercomputing extensions. An example of such a system is the HITAC S-3800, which was instruction-set compatible with IBM System/370 mainframes, and could run the Hitachi VOS3 operating system (a fork of IBM MVS). The S-3800 therefore can be seen as being both simultaneously a supercomputer and also an IBM-compatible mainframe. In 2007, an amalgamation of the different technologies and architectures for supercomputers and mainframes has led to a so-called gameframe.

Command-Line Interfaces (CLIs)

A command-line interface (CLI) is a means of interacting with a computer program by inputting lines of text called command-lines. Command-line interfaces emerged in the mid-1960s, on computer terminals, as an interactive and more user-friendly alternative to the non-interactive interface available with punched cards. Today, most computer users rely on graphical user interfaces ("GUIs") instead of CLIs. However, many programs and operating system utilities lack GUIs, and are intended to be used through CLIs. Knowledge of CLIs is also useful for writing scripts. Programs that have CLIs are generally easy to automate via scripting, since command-lines, being mere lines of text, are easy to specify in code.

CLIs are made possible by command-line interpreters or command-line processors, which are programs that read command-lines and carry out the commands. Alternatives to CLIs include GUIs (most notably desktop metaphors with a mouse pointer, such as Microsoft Windows), text-based user interface menus (such as DOS Shell and IBM AIX SMIT), and keyboard shortcuts.

Comparison to Graphical User Interfaces

Compared with a graphical user interface, a command-line interface requires fewer system resources to implement. Since options to commands are given in a few characters in each command line, an experienced user often finds the options easier to access. Automation of repetitive tasks is simplified by line editing and history mechanisms for storing frequently used sequences; this may extend to a scripting language that can take parameters and variable options. A command-line history can be kept, allowing review or repetition of commands.

A command-line system may require paper or online manuals for the user's reference, although often a "help" option provides a concise review of the options of a command. The command-line environment may not provide graphical enhancements such as different fonts or extended edit windows found in a GUI. It may be difficult for a new user to become familiar with all the commands and options available, compared with the icons and drop-down menus of a graphical user interface, without reference to manuals.

Operating System Command-Line Interfaces

Operating system (OS) command-line interfaces are usually distinct programs supplied with the operating system. A program that implements such a text interface is often called a command-line interpreter, command processor or shell.

Examples of command-line interpreters include DEC's DIGITAL Command Language (DCL) in OpenVMS and RSX-11, the various Unix shells (sh, ksh, csh, tcsh, zsh, Bash, etc.), CP/M's CCP, DOS' COMMAND.COM, as well as the OS/2 and the Windows CMD.EXE programs, the latter groups being based heavily on DEC's RSX-11 and

RSTS CLIs. Under most operating systems, it is possible to replace the default shell program with alternatives; examples include 4DOS for DOS, 4OS2 for OS/2, and 4NT / Take Command for Windows.

Although the term 'shell' is often used to describe a command-line interpreter, strictly speaking, a 'shell' can be any program that constitutes the user-interface, including fully graphically oriented ones. For example, the default Windows GUI is a shell program named EXPLORER.EXE, as defined in the SHELL=EXPLORER.EXE line in the WIN.INI configuration file. These programs are shells, but not CLIs.

Application Command-Line Interfaces

Application programs (as opposed to operating systems) may also have command-line interfaces. An application program may support none, any, or all of these three major types of command-line interface mechanisms:

- **Parameters:** Most command-line interfaces support a means to pass additional information to a program when it is launched.
- **Interactive Command-Line Sessions:** After launch, a program may provide an operator with an independent means to enter commands.
- **Inter-Process Communication:** Most operating systems support means of inter-process communication (for example, standard streams or named pipes). Command lines from client processes may be redirected to a CLI program by one of these methods.

Some applications support a CLI, presenting their own prompt to the user and accepting command lines. Other programs support both a CLI and a GUI. In some cases, a GUI is simply a wrapper around a separate CLI executable file. In other cases, a program may provide a CLI as an optional alternative to its GUI. CLIs and GUIs often support different functionality. For example, all features of MATLAB, a numerical analysis computer program, are available via the CLI, whereas the MATLAB GUI exposes only a subset of features. In Colossal Cave Adventure from 1975, the user uses a CLI to enter one or two words to explore a cave system.

The Backdrop

The command-line interface evolved from a form of communication conducted by people over teleprinter (TTY) machines. Sometimes these involved sending an order or a confirmation using telex. Early computer systems often used teleprinter as the means of interaction with an operator.

The mechanical teleprinter was replaced by a "glass tty", a keyboard and screen emulating the teleprinter. "Smart" terminals permitted additional functions, such as cursor movement over the entire screen, or local editing of data on the terminal for transmission to the computer. As the microcomputer revolution replaced the traditional – minicomputer + terminals – time sharing architecture, hardware terminals were replaced by terminal emulators — PC software that interpreted terminal signals sent through the PC's serial ports. These were typically used to interface an organization's new PC's with their existing mini- or mainframe computers, or to connect PC to PC. Some of these PCs were running Bulletin Board System software.

Early operating system CLIs were implemented as part of resident monitor programs, and could not easily be replaced. The first implementation of the shell as a replaceable component was part of the Multics time-sharing operating system. In 1964, MIT Computation Center staff member Louis Pouzin developed the RUNCOM tool for executing command scripts while allowing argument substitution. Pouzin coined the term "shell" to describe the technique of using commands like a programming language, and wrote a paper about how to implement the idea in the Multics operating system. Pouzin returned to his native France in 1965, and the first Multics shell was developed by Glenda Schroeder.

The first Unix shell, the V6 shell, was developed by Ken Thompson in 1971 at Bell Labs and was modeled after Schroeder's Multics shell. The Bourne shell was introduced in 1977 as a replacement for the V6 shell. Although it is used as an interactive command interpreter, it was also intended as a scripting language and contains most of the features that are commonly considered to produce structured programs. The Bourne shell led to the development of the KornShell (ksh), Almquist shell (ash), and the popular Bourne-again shell (or Bash).

Early microcomputers themselves were based on a command-line interface such as CP/M, DOS or ApplcSoft BASIC. During the 1980s and 1990s, the introduction of the Apple Macintosh and of Microsoft Windows on PCs saw the command line interface as the primary user interface replaced by the Graphical User Interface. The command line remained available as an alternative user interface, often used by system administrators and other advanced users for system administration, computer programming and batch processing.

In November 2006, Microsoft released version 1.0 of Windows PowerShell (formerly codenamed *Monad*), which combined features of traditional Unix shells with their proprietary object-oriented .NET Framework. MinGW and Cygwin are open-source packages for Windows that offer a Unix-like CLI. Microsoft provides MKS Inc.'s ksh implementation *MKS Korn shell* for Windows through their Services for UNIX add-on.

Since 2001, the Macintosh operating system macOS has been based on a Unix-like operating system called Darwin. On these computers, users can access a Unix-like command-line interface by running the terminal emulator program called Terminal, which is found in the Utilities sub-folder of the Applications folder, or by remotely logging into the machine using ssh. Z shell is the default shell for macOS; Bash, tcsh, and the KornShell are also provided. Before macOS Catalina, Bash was the default.

Usage

A CLI is used whenever a large vocabulary of commands or queries, coupled with a wide (or arbitrary) range of options, can be entered more rapidly as text than with a pure GUI. This is typically the case with operating system command shells. CLIs are also used by systems with insufficient resources to support a graphical user interface. Some computer language systems (such as Python, Forth, LISP, Rexx, and many dialects of BASIC) provide an interactive command-line mode to allow for rapid evaluation of code.

CLIs are often used by programmers and system administrators, in engineering and scientific environments, and by technically advanced personal computer users. CLIs are also popular among people with visual disabilities since the commands and responses can be displayed using refreshable Braille displays.

Command-Line Interpreter

- Although most users think of the shell as an interactive command interpreter, it is really a programming language in which each statement runs a command. Because it must satisfy both the interactive and programming aspects of command execution, it is a strange language, shaped as much by history as by design.—*Brian W. Kernighan & Rob Pike*

The term command-line interpreter (CLI) is applied to computer programs designed to interpret a sequence of lines of text which may be entered by a user, read from a file or another kind of data stream. The context of interpretation is usually one of a given operating system or programming language.

Command-line interpreters allow users to issue various commands in a very efficient (and often terse) way. This requires the user to know the names of the commands and their parameters, and the syntax of the language that is interpreted.

The Unix #! mechanism and OS/2 EXTPROC command facilitate the passing of batch files to external processors. One can use these mechanisms to write specific command processors for dedicated uses, and process external data files which reside in batch files.

Many graphical interfaces, such as the OS/2 Presentation Manager and early versions of Microsoft Windows use command-lines to call helper programs to open documents and programs. The commands are stored in the graphical shell or in files like the registry or the OS/2 OS2USER.INI file. The earliest computers did not support interactive input/output devices, often relying on sense switches and lights to communicate with the computer operator. This was adequate for batch systems that ran one program at a time, often with the programmer acting as operator.

This also had the advantage of low overhead, since lights and switches could be tested and set with one machine instruction. Later a single system console was added to allow the operator to communicate with the system.

From the 1960s onwards, user interaction with computers was primarily by means of command-line interfaces, initially on machines like the Teletype Model 33 ASR, but then on early CRT-based computer terminals such as the VT52.

All of these devices were purely text based, with no ability to display graphic or pictures. For business application programs, text-based menus were used, but for more general interaction the command line was the interface.

Around 1964 Louis Pouzin introduced the concept and the name shell in Multics, building on earlier, simpler facilities in the Compatible Time-Sharing System (CTSS).

From the early 1970s the Unix operating system adapted the concept of a powerful command-line environment, and introduced the ability to pipe the output of one command in as input to another. Unix also had the capability to save and re-run strings of commands as "shell scripts" which acted like custom commands.

The command-line was also the main interface for the early home computers such as the Commodore PET, Apple II and BBC Micro – almost always in the form of a BASIC interpreter. When more powerful business oriented microcomputers arrived with CP/M and later DOS computers such as the IBM PC, the command-line began to borrow some of the syntax and features of the Unix shells such as globbing and piping of output.

The command-line was first seriously challenged by the PARC GUI approach used in the 1983 Apple Lisa and the 1984 Apple Macintosh. A few computer users used GUIs such as GEOS and Windows 3.1 but the majority of IBM PC users did not replace their COMMAND.COM shell with a GUI until Windows 95 was released in 1995.

While most non-expert computer users now use a GUI almost exclusively, more advanced users have access to powerful command-line environments:

- The default VAX/VMS command shell, using the DCL language, has been ported to Windows systems at least three times, including PC-DCL and Acceler8 DCL Lite. Unix command shells have been ported to VMS and DOS/Windows 95 and Windows NT types of operating systems.
- COMMAND.COM is the command-line interpreter of MS-DOS, IBM PC DOS, and clones such as DR-DOS, SISNE plus, PTS-DOS, ROM-DOS, and FreeDOS.

- Windows Resource Kit and Windows Services for UNIX include Korn and the Bourne shells along with a Perl interpreter (Services for UNIX contains ActiveState ActivePerl in later versions and Interix for versions 1 and 2 and a shell compiled by Microsoft)
- IBM OS/2 (and derivatives such as eComStation and ArcaOS) has the cmd.exe processor. This copies the COMMAND.COM commands, with extensions to REXX.
- cmd.exe is part of the Windows NT stream of operating systems.
- Yet another cmd.exe is a stripped-down shell for Windows CE 3.0.
- An MS-DOS type interpreter called PocketDOS has been ported to Windows CE machines; the most recent release is almost identical to MS-DOS 6.22 and can also run Windows 1, 2, and 3.0, QBasic and other development tools, 4NT and 4DOS. The latest release includes several shells, namely MS-DOS 6.22, PC DOS 7, DR DOS 3.xx, and others.
- Windows users might use the CScript interface to alternate programs, from command-line. PowerShell provides a command-line interface, but its applets are not written in Shell script. Implementations of the Unix shell are also available as part of the POSIX sub-system, Cygwin, MKS Toolkit, UWIN, Hamilton C shell and other software packages. Available shells for these interoperability tools include csh, ksh, sh, Bash, rsh, tclsh and less commonly zsh, psh
- Implementations of PHP have a shell for interactive use called php-cli.
- Standard Tcl/Tk has two interactive shells, Tclsh and Wish, the latter being the GUI version.
- Python, Ruby, Lua, XLNT, and other interpreters also have command shells for interactive use.
- FreeBSD uses tcsh as its default interactive shell for the superuser, and ash as default scripting shell.
- Many Linux distributions have the Bash implementation of the Unix shell.
- Apple macOS and some Linux distributions use zsh. Previously, macOS used tcsh and Bash.
- Embedded Linux (and other embedded Unix-like) devices often use the Ash implementation of the Unix shell, as part of Busybox.
- Android uses the mksh shell, which replaces a shell derived from ash that was used in older Android versions, supplemented with commands from the separate *toolbox* binary.

- Routers with Cisco IOS, Junos and many others are commonly configured from the command line.
- The Plan 9 operating system uses the rc shell which is similar in design to the Bourne shell.

Scripting

Most command-line interpreters support scripting, to various extents. (They are, after all, interpreters of an interpreted programming language, albeit in many cases the language is unique to the particular command-line interpreter.) They will interpret scripts (variously termed shell scripts or batch files) written in the language that they interpret. Some command-line interpreters also incorporate the interpreter engines of other languages, such as REXX, in addition to their own, allowing the executing of scripts, in those languages, directly within the command-line interpreter itself.

Conversely, scripting programming languages, in particular those with an eval function (such as REXX, Perl, Python, Ruby or Jython), can be used to implement command-line interpreters and filters. For a few operating systems, most notably DOS, such a command interpreter provides a more flexible command-line interface than the one supplied. In other cases, such a command interpreter can present a highly customised user interface employing the user interface and input/output facilities of the language.

Other Command-Line Interfaces

The command line provides an interface between programs as well as the user. In this sense, a command line is an alternative to a dialog box. Editors and databases present a command line, in which alternate command processors might run. On the other hand, one might have options on the command line, which opens a dialog box. The latest version of 'Take Command' has this feature. DBase used a dialog box to construct command lines, which could be further edited before use.

Programs like BASIC, diskpart, Edlin, and QBASIC all provide command-line interfaces, some of which use the system shell. Basic is modeled on the default interface for 8-bit Intel computers. Calculators can be run as command-line or dialog interfaces.

Emacs provides a command-line interface in the form of its minibuffer. Commands and arguments can be entered using Emacs standard text editing support, and output is displayed in another buffer.

There are a number of text mode games, like *Adventure* or *King's Quest 1-3*, which relied on the user typing commands at the bottom of the screen. One controls the character by typing commands like 'get ring' or 'look'. The program returns a text which describes how the character sees it, or makes the action happen. The text adventure *The Hitchhiker's Guide to the Galaxy*, a piece of interactive fiction based on Douglas Adam's book of the same name, is a teletype-style command-line game.

The most notable of these interfaces is the standard streams interface, which allows the output of one command to be passed to the input of another. Text files can serve either purpose as well. This provides the interfaces of piping, filters and redirection. Under Unix, devices are files too, so the normal type of file for the shell used for stdin, stdout and stderr is a tty device file.

Another command-line interface allows a shell program to launch helper programs, either to launch documents or start a program. The command is processed internally by the shell, and then passed on to another program to launch the document. The graphical interface of Windows and OS/2 rely heavily on command-lines passed through to other programs – console or graphical, which then usually process the command line without presenting a user-console.

Programs like the OS/2 E editor and some other IBM editors, can process command-lines normally meant for the shell, the output being placed directly in the document window.

A web browser's URL input field can be used as a command line. It can be used to "launch" web apps, access browser configuration, as well as perform a search. Google, which has been called "the command line of the internet" will perform a domain-specific search when it detects search parameters in a known format. This functionality is present whether the search is triggered from a browser field or on Google's website.

There are JavaScript libraries that allow to write command line applications in browser as standalone Web apps or as part of bigger application. An example of such a website is the CLI interface to DuckDuckGo. There are also Web-based SSH applications, that allow to give access to server command line interface from a browser.

Many PC video games feature a command line interface often referred to as a console. It is typically used by the game developers during development and by mod developers for debugging purposes as well as for cheating or skipping parts of the game.

1.1.3 The Personal Computer Revolution

Graphical User Interfaces (GUIs)

A graphical user interface, or GUI is a form of user interface that allows users to interact with electronic devices through graphical icons and visual indicators such as secondary notation. In many applications, GUIs are used instead of text-based UIs, which are based on typed command labels or text navigation. GUIs were introduced in reaction to the perceived steep learning curve of command-line interfaces (CLIs), which require commands to be typed on a computer keyboard.

The actions in a GUI are usually performed through direct manipulation of the graphical elements. Beyond computers, GUIs are used in many handheld mobile devices such as MP3 players, portable media players, gaming devices, smartphones and smaller household, office and industrial controls. The term *GUI* tends not to be applied to other lower-display resolution types of interfaces, such as video games (where head-up displays (*HUDs*) are preferred), or not including flat screens like volumetric displays because the term is restricted to the scope of 2D display screens able to describe generic information, in the tradition of the computer science research at the Xerox Palo Alto Research Center.

GUI and Interaction Design

Designing the visual composition and temporal behavior of a GUI is an important part of software application programming in the area of human–computer interaction. Its goal is to enhance the efficiency and ease of use for the underlying logical design of a stored program, a design discipline named *usability*. Methods of user-centered design are used to ensure that the visual language introduced in the design is well-tailored to the tasks.

The visible graphical interface features of an application are sometimes referred to as *chrome* or *GUI*. Typically, users interact with information by manipulating visual widgets that allow for interactions appropriate to the kind of data they hold. The widgets of a well-designed interface are selected to support the actions necessary to achieve the goals of users. A model–view–controller allows flexible structures in which the interface is independent of and indirectly linked to application functions, so the GUI can be customized easily.

This allows users to select or design a different *skin* or *theme* at will, and eases the designer's work to change the interface as user needs evolve. Good GUI design relates to users more, and to system architecture less. Large widgets, such as windows, usually provide a frame or container for the main presentation content such as a web page, email message, or drawing. Smaller ones usually act as a user-input tool.

A GUI may be designed for the requirements of a vertical market as application-specific GUIs. Examples include automated teller machines (ATM), point of sale (POS) touchscreens at restaurants, self-service checkouts used in a retail store, airline self-ticket and check-in, information kiosks in a public space, like a train station or a museum, and monitors or control screens in an embedded industrial application which employ a real-time operating system (RTOS).

Cell phones and handheld game systems also employ application specific touchscreen GUIs. Newer automobiles use GUIs in their navigation systems and multimedia centers, or navigation multimedia center combinations.

Components

A GUI uses a combination of technologies and devices to provide a platform that users can interact with, for the tasks of gathering and producing information. A series of elements conforming a visual language have evolved to represent information stored in computers. This makes it easier for people with few computer skills to work with and use computer software. The most common combination of such elements in GUIs is the *windows, icons, text fields, canvases, menus, pointer* (WIMP) paradigm, especially in personal computers.

The WIMP style of interaction uses a virtual input device to represent the position of a pointing device's interface, most often a mouse, and presents information organized in windows and represented with icons. Available commands are compiled together in menus, and actions are performed making gestures with the pointing device. A window manager facilitates the interactions between windows, applications, and the windowing system. The windowing system handles hardware devices such as pointing devices, graphics hardware, and positioning of the pointer.

In personal computers, all these elements are modeled through a desktop metaphor to produce a simulation called a desktop environment in which the display represents a desktop, on which documents and folders of documents can be placed. Window managers and other software combine to simulate the desktop environment with varying degrees of realism.

Entries may appear in a list to make space for text and details, or in a grid for compactness and larger icons with little space underneath for text. Variations inbetween exist, such as a list with multiple columns of items and a grid of items with rows of text extending sideways from the icon.

Multi-row and multi-column layouts commonly found on the web are "shelf" and "waterfall". The former is found on image search engines, where images appear with a fixed height but variable length, and is typically implemented with the CSS property and parameter display: inline-block;. A waterfall layout found on Imgur and TweetDeck with fixed width but variable height per item is usually implemented by specifying column-width:.

Post-WIMP Interface

Smaller app mobile devices such as personal digital assistants (PDAs) and smartphones typically use the WIMP elements with different unifying metaphors, due to constraints in space and available input devices. Applications for which WIMP is not well suited may use newer interaction techniques, collectively termed *post-WIMP* UIs. As of 2011, some touchscreen-based operating systems such as Apple's iOS (iPhone) and Android use the class of GUIs named post-WIMP. These support styles of interaction using more than one finger in contact with a display, which allows actions such as pinching and rotating, which are unsupported by one pointer and mouse.

Interaction

Human interface devices, for the efficient interaction with a GUI include a computer keyboard, especially used together with keyboard shortcuts, pointing devices for the cursor (or rather pointer) control: mouse, pointing stick, touchpad, trackball, joystick, virtual keyboards, and head-up displays (translucent information devices at the eye level). There are also actions performed by programs that affect the GUI. For example, there are components like inotify or D-Bus to facilitate communication between computer programs.

The Backdrop

Ivan Sutherland developed Sketchpad in 1963, widely held as the first graphical computer-aided design program. It used a light pen to create and manipulate objects in engineering drawings in realtime with coordinated graphics. In the late 1960s, researchers at the Stanford Research Institute, led by Douglas Engelbart, developed the On-Line System (NLS), which used text-based hyperlinks manipulated with a then-new device: the mouse. (A 1968 demonstration of NLS became known as "The Mother of All Demos".) In the 1970s, Engelbart's ideas were further refined and extended to graphics by researchers at Xerox PARC and specifically Alan Kay, who went beyond text-based hyperlinks and used a GUI as the main interface for the Smalltalk programming language, which ran on the Xerox Alto computer, released in 1973. Most modern general-purpose GUIs are derived from this system.

The Xerox PARC GUI consisted of graphical elements such as windows, menus, radio buttons, and check boxes. The concept of icons was later introduced by David Canfield Smith, who had written a thesis on the subject under the guidance of Kay. The PARC GUI employs a pointing device along with a keyboard. These aspects can be emphasized by using the alternative term and acronym for *windows, icons, menus, pointing device* (WIMP). This effort culminated in the 1973 Xerox Alto, the first computer with a GUI, though the system never reached commercial production.

The first commercially available computer with a GUI was the 1979 PERQ workstation, manufactured by Three Rivers Computer Corporation. Its design was heavily influenced by the work at Xerox PARC. In 1981, Xerox eventually commercialized the ideas from the Alto in the form of a new and enhanced system – the Xerox 8010 Information System – more commonly known as the Xerox Star. These early systems spurred many other GUI efforts, including Lisp machines by

Symbolics and other manufacturers, the Apple Lisa (which presented the concept of menu bar and window controls) in 1983, the Apple Macintosh 128K in 1984, and the Atari ST with Digital Research's GEM, and Commodore Amiga in 1985. Visi On was released in 1983 for the IBM PC compatible computers, but was never popular due to its high hardware demands. Nevertheless, it was a crucial influence on the contemporary development of Microsoft Windows.

Apple, Digital Research, IBM and Microsoft used many of Xerox's ideas to develop products, and IBM's Common User Access specifications formed the basis of the GUIs used in Microsoft Windows, IBM OS/2 Presentation Manager, and the Unix Motif toolkit and window manager. These ideas evolved to create the interface found in current versions of Microsoft Windows, and in various desktop environments for Unix-like operating systems, such as macOS and Linux. Thus most current GUIs have largely common idioms.

GUIs were a hot topic in the early 1980s. The Apple Lisa was released in 1983, and various windowing systems existed for DOS operating systems (including PC GEM and PC/GEOS). Individual applications for many platforms presented their own GUI variants. Despite the GUIs advantages, many reviewers questioned the value of the entire concept, citing hardware limits, and problems in finding compatible software.

In 1984, Apple released a television commercial which introduced the Apple Macintosh during the telecast of Super Bowl XVIII by CBS, with allusions to George Orwell's noted novel *Nineteen Eighty-Four*. The goal of the commercial was to make people think about computers, identifying the user-friendly interface as a personal computer which departed from prior business-oriented systems, and becoming a signature representation of Apple products.

In 1985, Commodore released the Amiga 1000, along with Workbench and Kickstart 1.0 (which contained Intuition). This interface ran as a separate task, meaning it was very responsive and, unlike other GUIs of the time, it didn't freeze up when a program was busy. Additionally, it was the first GUI to introduce something resembling Virtual Desktops.

Windows 95, accompanied by an extensive marketing campaign, was a major success in the marketplace at launch and shortly became the most popular desktop operating system.

In 2007, with the iPhone and later in 2010 with the introduction of the iPad, Apple popularized the post-WIMP style of interaction for multi-touch screens, and those devices were considered to be milestones in the development of mobile devices.

The GUIs familiar to most people as of the mid-late 2010s are Microsoft Windows, macOS, and the X Window System interfaces for desktop and laptop computers, and Android, Apple's iOS, Symbian, BlackBerry OS, Windows Phone/Windows 10 Mobile, Tizen, WebOS, and Firefox OS for handheld (smartphone) devices.

Comparison to other Interfaces

Since the commands available in command line interfaces can be many, complex operations can be performed using a short sequence of words and symbols. Custom functions may be used to facilitate access to frequent actions. Command-line interfaces are more lightweight, as they only recall information necessary for a task; for example, no preview thumbnails or graphical rendering of web pages. This allows greater efficiency and productivity once many commands are learned. But reaching this level takes some time because the command words may not be easily discoverable or mnemonic.

Also, using the command line can become slow and error-prone when users must enter long commands comprising many parameters or several different filenames at once. However, *windows, icons, menus, pointer* (WIMP) interfaces present users with many widgets that represent and can trigger some of the system's available commands.

GUIs can be made quite hard when dialogs are buried deep in a system or moved about to different places during redesigns. Also, icons and dialog boxes are usually harder for users to script. WIMPs extensively use modes, as the meaning of all keys and clicks on specific positions on the screen are redefined all the time. Command-line interfaces use modes only in limited forms, such as for current directory and environment variables.

Most modern operating systems provide both a GUI and some level of a CLI, although the GUIs usually receive more attention. GUI wrappers find a way around the command-line interface versions (CLI) of (typically) Linux and Unix-like software applications and their text-based UIs or typed command labels. While command-line or text-based applications allow users to run a program non-interactively, GUI wrappers atop them avoid the steep learning curve of the command-line, which requires commands to be typed on the keyboard.

By starting a GUI wrapper, users can intuitively interact with, start, stop, and change its working parameters, through graphical icons and visual indicators of a desktop environment, for example. Applications may also provide both interfaces, and when they do the GUI is usually a WIMP wrapper around the command-line version. This is especially common with applications designed for Unix-like operating systems. The latter used to be implemented first because it allowed the developers to focus exclusively on their product's functionality without bothering about interface details such as designing icons and placing buttons. Designing programs this way also allows users to run the program in a shell script.

Three-Dimensional Graphical User Interface

Many environments and games use the methods of 3D graphics to project 3D GUI objects onto the screen. The use of 3D graphics has become increasingly common in mainstream operating systems (ex. Windows Aero, and Aqua (MacOS)) to create attractive interfaces, termed eye candy (which includes, for example, the use of drop shadows underneath windows and the cursor), or for functional purposes only possible using three dimensions. For example, user switching is represented by rotating a cube with faces representing each user's workspace, and window management is represented via a Rolodex-style flipping mechanism in Windows Vista. In both cases, the operating system transforms windows on-the-fly while continuing to update the content of those windows.

The GUI is usually WIMP-based, although occasionally other metaphors surface, such as those used in Microsoft Bob, 3dwm, File System Navigator, File System Visualizer, 3D Mailbox, and GopherVR. Zooming (ZUI) is a related technology that promises to deliver the representation benefits of 3D environments without their usability drawbacks of orientation problems and hidden objects. In 2006, Hillcrest Labs introduced the first ZUI for television. Other innovations include the menus on the PlayStation 2, the menus on the Xbox, Sun's Project Looking Glass, Metisse, which was similar to Project Looking Glass, BumpTop, where users can manipulate documents and windows with realistic movement and physics as if they were physical documents, Croquet OS, which is built for collaboration, and compositing window managers such as Enlightenment and Compiz. Augmented reality and virtual reality also make use of 3D GUI elements.

The Mouse and Keyboard

The mouse and keyboard became the standard input devices during this period, replacing command-line interaction with point-and-click simplicity. These devices enabled users to interact with on-screen elements directly, fostering a more engaging and productive computing experience. The combination of GUIs and these input devices established a new standard for HMI, one that remains influential to this day.

Computer Mouse

A computer mouse is a hand-held pointing device that detects two-dimensional motion relative to a surface. This motion is typically translated into the motion of the pointer (called a cursor) on a display, which allows a smooth control of the graphical user interface of a computer.

The first public demonstration of a mouse controlling a computer system was done by Doug Engelbart in 1968 as part of the Mother of All Demos. Mice originally used two separate wheels to directly track movement across a surface: one in the x-dimension and one in the Y. Later, the standard design shifted to use a ball rolling on a surface to detect motion, in turn connected to internal rollers. Most modern mice use optical movement detection with no moving parts. Though originally all mice were connected to a computer by a cable, many modern mice are cordless, relying on short-range radio communication with the connected system.

In addition to moving a cursor, computer mice have one or more buttons to allow operations such as the selection of a menu item on a display. Mice often also feature other elements, such as touch surfaces and scroll wheels, which enable additional control and dimensional input.

History

The trackball, a related pointing device, was invented in 1946 by Ralph Benjamin as part of a post-World War II-era fire-control radar plotting system called the Comprehensive Display System (CDS). Benjamin was then working for the British Royal Navy Scientific Service. Benjamin's project used analog computers to calculate the future position of target aircraft based on several initial input points provided by a user with a joystick. Benjamin felt that a more elegant input device was needed and invented what they called a "roller ball" for this purpose.

The device was patented in 1947, but only a prototype using a metal ball rolling on two rubber-coated wheels was ever built, and the device was kept as a military secret.

Another early trackball was built by Kenyon Taylor, a British electrical engineer working in collaboration with Tom Cranston and Fred Longstaff. Taylor was part of the original Ferranti Canada, working on the Royal Canadian Navy's DATAR (Digital Automated Tracking and Resolving) system in 1952.

DATAR was similar in concept to Benjamin's display. The trackball used four disks to pick up motion, two each for the X and Y directions. Several rollers provided mechanical support. When the ball was rolled, the pickup discs spun and contacts on their outer rim made periodic contact with wires, producing pulses of output with each movement of the ball.

By counting the pulses, the physical movement of the ball could be determined. A digital computer calculated the tracks and sent the resulting data to other ships in a task force using pulse-code modulation radio signals. This trackball used a standard Canadian five-pin bowling ball. It was not patented, since it was a secret military project.

Douglas Engelbart of the Stanford Research Institute (now SRI International) has been credited in published books by Thierry Bardini, Paul Ceruzzi, Howard Rheingold, and several others as the inventor of the computer mouse. Engelbart was also recognized as such in various obituary titles after his death in July 2013.

By 1963, Engelbart had already established a research lab at SRI, the Augmentation Research Center (ARC), to pursue his objective of developing both hardware and software computer technology to "augment" human intelligence. That November, while attending a conference on computer graphics in Reno, Nevada, Engelbart began to ponder how to adapt the underlying principles of the planimeter to inputting X- and Y-coordinate data. On 14 November 1963, he first recorded his thoughts in his personal notebook about something he initially called a "bug", which is a "3-point" form could have a "drop point and 2 orthogonal wheels". He wrote that the "bug" would be "easier" and "more natural" to use, and unlike a stylus, it would stay still when let go, which meant it would be "much better for coordination with the keyboard".

In 1964, Bill English joined ARC, where he helped Engelbart build the first mouse prototype. They christened the device the *mouse* as early models had a cord attached to the rear part of the device which looked like a tail, and in turn, resembled the common mouse. According to Roger Bates, a hardware designer in English, another reason for choosing this name was because the cursor on the screen was also referred to as "CAT" at this time.

As noted above, this "mouse" was first mentioned in print in a July 1965 report, on which English was the lead author. On 9 December 1968, Engelbart publicly demonstrated the mouse at what would come to be known as The Mother of All Demos. Engelbart never received any royalties for it, as his employer SRI held the patent, which expired before the mouse became widely used in personal computers. In any event, the invention of the mouse was just a small part of Engelbart's much larger project of augmenting human intellect.

Several other experimental pointing-devices developed for Engelbart's oN-Line System (NLS) exploited different body movements – for example, head-mounted devices attached to the chin or nose – but ultimately the mouse won out because of its speed and convenience. The first mouse, a bulky device (pictured) used two potentiometers perpendicular to each other and connected to wheels: the rotation of each wheel translated into motion along one axis. At the time of the "Mother of All Demos", Engelbart's group had been using their second-generation, 3-button mouse for about a year.

On 2 October 1968, three years after Engelbart's prototype but more than two months before his public demo, a mouse device named *Rollkugelsteuerung* (German for "Trackball control") was shown in a sales brochure by the German company AEG-Telefunken as an optional input device for the SIG 100 vector graphics terminal, part of the system around their process computer TR 86 and the TR 440 [de] main frame. Based on an even earlier trackball device, the mouse device had been developed by the company in 1966 in what had been a parallel and independent discovery. As the name suggests and unlike Engelbart's mouse, the Telefunken model already had a ball (diameter 40 mm, weight 40 g) and two mechanical 4-bit rotational position transducers with Gray code-like states, allowing easy movement in any direction. The bits remained stable for at least two successive states to relax debouncing requirements. This arrangement was chosen so that the data could also be transmitted to the TR 86 front-end process computer and over longer distance telex lines with c. 50 baud. Weighing 465 grams (16.4 oz), the device with a total height of about 7 cm (2.8 in) came in a c. 12 cm (4.7 in) diameter hemispherical injection-molded thermoplastic casing featuring one central push button.

As noted above, the device was based on an earlier trackball-like device (also named *Rollkugel*) that was embedded into radar flight control desks. This trackball had been originally developed by a team led by Rainer Mallebrein [de] at Telefunken *Konstanz* for the German *Bundesanstalt für Flugsicherung* [de] (Federal Air Traffic Control). It was part of the corresponding workstation system SAP 300 and the terminal SIG 3001, which had been designed and developed since 1963. Development for the TR 440 main frame began in 1965. This led to the development of the TR 86 process computer system with its SIG 100-86 terminal. Inspired by a discussion with a university customer, Mallebrein came up with the idea of "reversing" the existing *Rollkugel* trackball into a moveable mouse-like device in 1966, so that customers did not have to be bothered with mounting holes for the earlier trackball device. The device was finished in early 1968, and together with light pens and trackballs, it was commercially offered as an optional input device for their system starting later that year. Not all customers opted to buy the device, which added costs of DM 1,500 per piece to the already up to 20-million DM deal for the main frame, of which only a total of 46 systems were sold or leased. They were installed at more than 20 German universities including RWTH Aachen, Technical University Berlin, University of Stuttgart and Konstanz. Several *Rollkugel* mice installed at the Leibniz Supercomputing Centre in Munich in 1972 are well preserved in a museum, two others survived in a museum at Stuttgart University, two in Hamburg, the one from Aachen at the Computer History Museum in the US, and yet another sample was recently donated to the Heinz Nixdorf MuseumsForum (HNF) in Paderborn. Anecdotal reports claim that Telefunken's attempt to patent the device was rejected by the German Patent Office due to lack of inventiveness. For the air traffic control system, the Mallebrein team had already developed a precursor to touch screens in form of an ultrasonic-curtain-based pointing device in front of the display. In 1970, they developed a device named "Touchinput-*Einrichtung*" ("touch input device") based on a conductively coated glass screen.

The Xerox Alto was one of the first computers designed for individual use in 1973 and is regarded as the first modern computer to use a mouse. Alan Kay designed the 16-by-16 mouse cursor icon with its left edge vertical and right edge 45-degrees so it displays well on the bitmap.Inspired by PARC's Alto, the Lilith, a computer which had been developed by a team around Niklaus Wirth at ETH Zürich between 1978 and 1980, provided a mouse as well. The third marketed version of an integrated mouse shipped as a part of a computer and intended for personal computer navigation came with the Xerox 8010 Star in 1981.

By 1982, the Xerox 8010 was probably the best-known computer with a mouse. The Sun-1 also came with a mouse, and the forthcoming Apple Lisa was rumored to use one, but the peripheral remained obscure;Jack Hawley of The Mouse House reported that one buyer for a large organization believed at first that his company sold lab mice. Hawley, who manufactured mice for Xerox, stated that "Practically, I have the market all to myself right now"; a Hawley mouse cost $415. In 1982, Logitech introduced the P4 Mouse at the Comdex trade show in Las Vegas, its first hardware mouse. That same year Microsoft made the decision to make the MS-DOS program Microsoft Word mouse-compatible, and developed the first PC-compatible mouse. The Microsoft Mouse shipped in 1983, thus beginning the Microsoft Hardware division of the company. However, the mouse remained relatively obscure until the appearance of the Macintosh 128K (which included an updated version of the single-button Lisa Mouse) in 1984, and of the Amiga 1000 and the Atari ST in 1985.

Operation

A mouse typically controls the motion of a pointer in two dimensions in a graphical user interface (GUI). The mouse turns movements of the hand backward and forward, left and right into equivalent electronic signals that in turn are used to move the pointer.

The relative movements of the mouse on the surface are applied to the position of the pointer on the screen, which signals the point where actions of the user take place, so hand movements are replicated by the pointer. Clicking or pointing (stopping movement while the cursor is within the bounds of an area) can select files, programs or actions from a list of names, or (in graphical interfaces) through small images called "icons" and other elements. For example, a text file might be represented by a picture of a paper notebook and clicking while the cursor points at this icon might cause a text editing program to open the file in a window.

Different ways of operating the mouse cause specific things to happen in the GUI:

- **Point:** Stop the motion of the pointer while it is inside the boundaries of what the user wants to interact with. This act of pointing is what the "pointer" and "pointing device" are named after. In web design lingo, pointing is referred to as "hovering". This usage spread to web programming and Android programming, and is now found in many contexts.

- **Click:** Pressing and releasing a button.
  - *(left) Single-Click:* Clicking the main button.
  - *(left) Double-Click:* Clicking the button two times in quick succession counts as a different gesture than two separate single clicks.
  - *(left) Triple-Click:* Clicking the button three times in quick succession counts as a different gesture than three separate single clicks. Triple clicks are far less common in traditional navigation.
  - *Right-Click:* Clicking the secondary button. In modern applications, this frequently opens a context menu.
  - *Middle-Click:* Clicking the tertiary button. In most cases, this is also the scroll wheel.
  - Clicking the fourth button.
  - Clicking the fifth button.
  - The USB standard defines up to 65535 distinct buttons for mice and other such devices, although in practice buttons above 3 are rarely implemented.
- **Drag:** Pressing and holding a button, and moving the mouse before releasing the button. This is frequently used to move or copy files or other objects via drag and drop; other uses include selecting text and drawing in graphics applications.
- Mouse button chording or chord clicking:
  - Clicking with more than one button simultaneously.
  - Clicking while simultaneously typing a letter on the keyboard.
  - Clicking and rolling the mouse wheel simultaneously.
  - Clicking while holding down a modifier key.
- **Moving the Pointer a Long Distance:** When a practical limit of mouse movement is reached, one lifts up the mouse, brings it to the opposite edge of the working area while it is held above the surface, and then lowers it back onto the working surface. This is often not necessary, because acceleration software detects fast movement, and moves the pointer significantly faster in proportion than for slow mouse motion.
- **Multi-Touch:** This method is similar to a multi-touch touchpad on a laptop with support for tap input for multiple fingers, the most famous example being the Apple Magic Mouse.

The Concept of Gestural Interfaces

Gestural interfaces have become an integral part of modern computing, allowing users to interact with their devices in a more intuitive and natural way. In addition to traditional pointing-and-clicking actions, users can now employ gestural inputs to issue commands or perform specific actions. These stylized motions of the mouse cursor, known as "gestures", have the potential to enhance user experience and streamline workflow.

Mouse Gestures in Action

To illustrate the concept of gestural interfaces, let's consider a drawing program as an example. In this scenario, a user can employ a gesture to delete a shape on the canvas. By rapidly moving the mouse cursor in an "x" motion over the shape, the user can trigger the command to delete the selected shape. This gesture-based interaction enables users to perform actions quickly and efficiently without relying solely on traditional input methods.

Challenges and Benefits of Gestural Interfaces

While gestural interfaces offer a more immersive and interactive user experience, they also present challenges. One of the primary difficulties lies in the requirement of finer motor control from users. Gestures demand precise movements, which can be more challenging for individuals with limited dexterity or those who are new to this mode of interaction.

However, despite these challenges, gestural interfaces have gained popularity due to their ability to simplify complex tasks and improve efficiency. Several gestural conventions have become widely adopted, making them more accessible to users. One such convention is the drag and drop gesture, which has become pervasive across various applications and platforms.

The Drag and Drop Gesture

The drag and drop gesture is a fundamental gestural convention that enables users to manipulate objects on the screen seamlessly. It involves a series of actions performed by the user:

- Pressing the mouse button while the cursor hovers over an interface object.
- Moving the cursor to a different location while holding the button down.
- Releasing the mouse button to complete the action.

This gesture allows users to transfer or rearrange objects effortlessly. For instance, a user can drag and drop a picture representing a file onto an image of a trash can, indicating the intention to delete the file. This intuitive and visual approach to interaction has become synonymous with organizing digital content and simplifying file management tasks.

Standard Semantic Gestures

In addition to the drag and drop gesture, several other semantic gestures have emerged as standard conventions within the gestural interface paradigm. These gestures serve specific purposes and contribute to a more intuitive user experience. Some of the notable semantic gestures include:

- **Crossing-Based Goal:** This gesture involves crossing a specific boundary or threshold on the screen to trigger an action or complete a task. For example, swiping across the screen to unlock a device or confirm a selection.
- **Menu Traversal:** Menu traversal gestures facilitate navigation through hierarchical menus or options. Users can perform gestures such as swiping or scrolling to explore different menu levels or activate specific commands.
- **Pointing:** Pointing gestures involve positioning the mouse cursor over an object or element to interact with it. This fundamental gesture enables users to select, click, or access contextual menus.
- **Mouseover (Pointing or Hovering):** Mouseover gestures occur when the cursor is positioned over an object without clicking. This action often triggers a visual change or displays additional information about the object, providing users with real-time feedback.

These standard semantic gestures, along with the drag and drop convention, form the building blocks of gestural interfaces, allowing users to interact with digital content using intuitive and natural movements.

At the end of 20th century, digitizer mice (puck) with magnifying glass was used with AutoCAD for the digitizations of blueprints.

Other uses of the mouse's input occur commonly in special application domains. In interactive three-dimensional graphics, the mouse's motion often translates directly into changes in the virtual objects' or camera's orientation. For example, in the first-person shooter genre of games, players usually employ the mouse to control the direction in which the virtual player's "head" faces: moving the mouse up will cause the player to look up, revealing the view above the player's head. A related function makes an image of an object rotate so that all sides can be examined. 3D design and animation software often modally chord many different combinations to allow objects and cameras to be rotated and moved through space with the few axes of movement mice can detect.

When mice have more than one button, the software may assign different functions to each button. Often, the primary (leftmost in a right-handed configuration) button on the mouse will select items, and the secondary (rightmost in a right-handed) button will bring up a menu of alternative actions applicable to that item. For example, on platforms with more than one button, the Mozilla web browser will follow a link in response to a primary button click, will bring up a contextual menu of alternative actions for that link in response to a secondary-button click, and will often open the link in a new tab or window in response to a click with the tertiary (middle) mouse button.

Mechanical Mice

The German company Telefunken published on their early ball mouse on 2 October 1968. Telefunken's mouse was sold as optional equipment for their computer systems. Bill English, builder of Engelbart's original mouse, created a ball mouse in 1972 while working for Xerox PARC.

The ball mouse replaced the external wheels with a single ball that could rotate in any direction. It came as part of the hardware package of the Xerox Alto computer. Perpendicular chopper wheels housed inside the mouse's body chopped beams of light on the way to light sensors, thus detecting in their turn the motion of the ball. This variant of the mouse resembled an inverted trackball and became the predominant form used with personal computers throughout the 1980s and 1990s. The Xerox PARC group also settled on the modern technique of using both hands to type on a full-size keyboard and grabbing the mouse when required.

The ball mouse has two freely rotating rollers. These are located 90 degrees apart. One roller detects the forward-backward motion of the mouse and the other the left-right motion. Opposite the two rollers is a third one (white, in the photo, at 45 degrees) that is spring-loaded to push the ball against the other two rollers. Each roller is on the same shaft as an encoder wheel that has slotted edges; the slots interrupt infrared light beams to generate electrical pulses that represent wheel movement. Each wheel's disc has a pair of light beams, located so that a given beam becomes interrupted or again starts to pass light freely when the other beam of the pair is about halfway between changes.

Simple logic circuits interpret the relative timing to indicate which direction the wheel is rotating. This incremental rotary encoder scheme is sometimes called quadrature encoding of the wheel rotation, as the two optical sensors produce signals that are in approximately quadrature phase. The mouse sends these signals to the computer system via the mouse cable, directly as logic signals in very old mice such as the Xerox mice, and via a data-formatting IC in modern mice. The driver software in the system converts the signals into motion of the mouse cursor along X and Y axes on the computer screen.

The ball is mostly steel, with a precision spherical rubber surface. The weight of the ball, given an appropriate working surface under the mouse, provides a reliable grip so the mouse's movement is transmitted accurately. Ball mice and wheel mice were manufactured for Xerox by Jack Hawley, doing business as The

Mouse House in Berkeley, California, starting in 1975. Based on another invention by Jack Hawley, proprietor of the Mouse House, Honeywell produced another type of mechanical mouse. Instead of a ball, it had two wheels rotating at off axes. Key Tronic later produced a similar product.

Modern computer mice took form at the École Polytechnique Fédérale de Lausanne (EPFL) under the inspiration of Professor Jean-Daniel Nicoud and at the hands of engineer and watchmaker André Guignard. This new design incorporated a single hard rubber mouseball and three buttons, and remained a common design until the mainstream adoption of the scroll-wheel mouse during the 1990s. In 1985, René Sommer added a microprocessor to Nicoud's and Guignard's design. Through this innovation, Sommer is credited with inventing a significant component of the mouse, which made it more "intelligent"; though optical mice from Mouse Systems had incorporated microprocessors by 1984.

Another type of mechanical mouse, the "analog mouse" (now generally regarded as obsolete), uses potentiometers rather than encoder wheels, and is typically designed to be plug compatible with an analog joystick. The "Color Mouse", originally marketed by RadioShack for their Color Computer (but also usable on MS-DOS machines equipped with analog joystick ports, provided the software accepted joystick input) was the best-known example.

Optical and Laser Mice

Early optical mice relied entirely on one or more light-emitting diodes (LEDs) and an imaging array of photodiodes to detect movement relative to the underlying surface, eschewing the internal moving parts a mechanical mouse uses in addition to its optics. A laser mouse is an optical mouse that uses coherent (laser) light.

The earliest optical mice detected movement on pre-printed mousepad surfaces, whereas the modern LED optical mouse works on most opaque diffuse surfaces; it is usually unable to detect movement on specular surfaces like polished stone. Laser diodes provide good resolution and precision, improving performance on opaque specular surfaces. Later, more surface-independent optical mice use an optoelectronic sensor (essentially, a tiny low-resolution video camera) to take successive images of the surface on which the mouse operates. Battery powered, wireless optical mice flash the LED intermittently to save power, and only glow steadily when movement is detected.

Inertial and Gyroscopic Mice

Often called "air mice" since they do not require a surface to operate, inertial mice use a tuning fork or other accelerometer (US Patent 4787051) to detect rotary movement for every axis supported. The most common models (manufactured by Logitech and Gyration) work using 2 degrees of rotational freedom and are insensitive to spatial translation. The user requires only small wrist rotations to move the cursor, reducing user fatigue or "gorilla arm".

Usually cordless, they often have a switch to deactivate the movement circuitry between use, allowing the user freedom of movement without affecting the cursor position. A patent for an inertial mouse claims that such mice consume less power than optically based mice, and offer increased sensitivity, reduced weight and increased ease-of-use. In combination with a wireless keyboard an inertial mouse can offer alternative ergonomic arrangements which do not require a flat work surface, potentially alleviating some types of repetitive motion injuries related to workstation posture.

3D Mice

A 3D mouse is a computer input device for viewport interaction with at least three degrees of freedom (DoF), e.g. in 3D computer graphics software for manipulating virtual objects, navigating in the viewport, defining camera paths, posing, and desktop motion capture. 3D mice can also be used as spatial controllers for video game interaction, e.g. SpaceOrb 360. To perform such different tasks the used transfer function and the device stiffness are essential for efficient interaction.

The virtual motion is connected to the 3D mouse control handle via a transfer function. Position control means that the virtual position and orientation is proportional to the mouse handle's deflection whereas velocity control means that translation and rotation velocity of the controlled object is proportional to the handle deflection. A further essential property of a transfer function is its interaction metaphor:

- **Object-in-Hand Metaphor:** An exterocentrical metaphor whereby the scene moves in correspondence with the input device. If the handle of the input device is twisted clockwise the scene rotates clockwise. If the handle is moved left the scene shifts left, and so on.
- **Camera-in-Hand Metaphor:** An egocentrical metaphor whereby the user's view is controlled by direct movement of a virtual camera. If the handle is twisted clockwise the scene rotates counter-clockwise. If the handle is moved left the scene shifts right, and so on.

Ware and Osborne performed an experiment investigating these metaphors whereby it was shown that there is no single best metaphor. For manipulation tasks, the object-in-hand metaphor was superior, whereas for navigation tasks the camera-in-hand metaphor was superior.

Zhai used and the following three categories for device stiffness:

- **Isotonic Input:** An input device with zero stiffness, that is, there is no self-centering effect.

- **Elastic Input:** A device with some stiffness, that is, the forces on the handle are proportional to the deflections.
- **Isometric Input:** An elastic input device with infinite stiffness, that is, the device handle does not allow any deflection but records force and torque.

Logitech 3D Mouse (1990) was the first ultrasonic mouse and is an example of an isotonic 3D mouse having six degrees of freedom (6DoF). Isotonic devices have also been developed with less than 6DoF, e.g. the Inspector at Technical University of Denmark (5DoF input).

Other examples of isotonic 3D mice are motion controllers, i.e. is a type of game controller that typically uses accelerometers to track motion. Motion tracking systems are also used for motion capture e.g. in the film industry, although that these tracking systems are not 3D mice in a strict sense, because motion capture only means recording 3D motion and not 3D interaction.

Early 3D mice for velocity control were almost ideally isometric, e.g. SpaceBall 1003, 2003, 3003, and a device developed at Deutsches Zentrum für Luft und Raumfahrt (DLR), cf. US patent US4589810A.

At DLR an elastic 6DoF sensor was developed that was used in Logitech's SpaceMouse and in the products of 3DConnexion. SpaceBall 4000 FLX has a maximum deflection of approximately 3 mm (0.12 in) at a maximum force of approximately 10N, that is, a stiffness of approximately 33 N/cm (19 lb_f/in). SpaceMouse has a maximum deflection of 1.5 mm (0.059 in) at a maximum force of 4.4 N (0.99 lb_f), that is, a stiffness of approximately 30 N/cm (17 lb_f/in).

Taking this development further, the softly elastic Sundinlabs SpaceCat was developed. SpaceCat has a maximum translational deflection of approximately 15 mm (0.59 in) and maximum rotational deflection of approximately 30° at a maximum force less than 2N, that is, a stiffness of approximately 1.3 N/cm (0.74 lb_f/in). With SpaceCat Sundin and Fjeld reviewed five comparative experiments performed with different device stiffness and transfer functions and performed a further study comparing 6DoF softly elastic position control with 6DoF stiffly elastic velocity control in a positioning task.

They concluded that for positioning tasks position control is to be preferred over velocity control. They could further conjecture the following two types of preferred 3D mouse usage:

- Positioning, manipulation, and docking using isotonic or softly elastic position control and an object-in-hand metaphor.
- Navigation using softly or stiffly elastic rate control and a camera-in-hand metaphor.

3DConnexion's 3D mice have been commercially successful over decades. They are used in combination with the conventional mouse for CAD. The Space Mouse is used to orient the target object or change the viewpoint with the non-dominant hand, whereas the dominant hand operates the computer mouse for conventional CAD GUI operation. This is a kind of space-multiplexed input where the 6 DoF input device acts as a graspable user interface that is always connected to the view port.

With force feedback the device stiffness can dynamically be adapted to the task just performed by the user, e.g. performing positioning tasks with less stiffness than navigation tasks.

Tactile Mice

In 2000, Logitech introduced a "tactile mouse" known as the "iFeel Mouse" developed by Immersion Corporation that contained a small actuator to enable the mouse to generate simulated physical sensations. Such a mouse can augment user-interfaces with haptic feedback, such as giving feedback when crossing a window boundary. To surf the internet by touch-enabled mouse was first developed in 1996 and first implemented commercially by the Wingman Force Feedback Mouse. It requires the user to be able to feel depth or hardness; this ability was realized with the first electrorheological tactile mice but never marketed.

Pucks

Tablet digitizers are sometimes used with accessories called pucks, devices which rely on absolute positioning, but can be configured for sufficiently mouse-like relative tracking that they are sometimes marketed as mice.

Ergonomic Mice

As the name suggests, this type of mouse is intended to provide optimum comfort and avoid injuries such as carpal tunnel syndrome, arthritis, and other repetitive strain injuries. It is designed to fit natural hand position and movements, to reduce discomfort.

When holding a typical mouse, the ulna and radius bones on the arm are crossed. Some designs attempt to place the palm more vertically, so the bones take more natural parallel position.

Increasing mouse height and angling the mouse topcase can improve wrist posture without negatively affecting performance. Some limit wrist movement, encouraging arm movement instead, that may be less precise but more optimal from the health point of view. A mouse may be angled from the thumb downward to the opposite side – this is known to reduce wrist pronation. However such optimizations make the mouse right or left hand specific, making more problematic to change the tired hand. *Time* has criticized manufacturers for offering few or no left-handed ergonomic mice: "Oftentimes I felt like I was dealing with someone who'd never actually met a left-handed person before." Another solution is a pointing bar device. The so-called *roller bar mouse* is positioned snugly in front of the keyboard, thus allowing bi-manual accessibility.

Gaming Mice

These mice are specifically designed for use in computer games. They typically employ a wider array of controls and buttons and have designs that differ radically from traditional mice. They may also have decorative monochrome or programmable RGB LED lighting. The additional buttons can often be used for changing the sensitivity of the mouse or they can be assigned (programmed) to macros (i.e., for opening a program or for use instead of a key combination). It is also common for game mice, especially those designed for use in real-time strategy games such as *StarCraft*, or in multiplayer online battle arena games such as League of Legends to have a relatively high sensitivity, measured in dots per inch (DPI), which can be as high as 25,600. DPI and CPI are the same values that refer to the mouse's sensitivity. DPI is a misnomer used in the gaming world, and many manufacturers use it to refer to CPI, counts per inch. Some advanced mice from gaming manufacturers also allow users to adjust the weight of the mouse by adding or subtracting weights to allow for easier control. Ergonomic quality is also an important factor in gaming mouse, as extended gameplay times may render further use of the mouse to be uncomfortable. Some mice have been designed to have adjustable features such as removable and/or elongated palm rests, horizontally adjustable thumb rests and pinky rests. Some mice may include several different rests with their products to ensure comfort for a wider range of target consumers. Gaming mice are held by gamers in three styles of grip:

- **Palm Grip:** The hand rests on the mouse, with extended fingers.
- **Claw Grip:** Palm rests on the mouse, bent fingers.
- **Finger-Tip Grip:** Bent fingers, palm does not touch the mouse.

Connectivity and Communication Protocols

To transmit their input, typical cabled mice use a thin electrical cord terminating in a standard connector, such as RS-232C, PS/2, ADB, or USB. Cordless mice instead transmit data via infrared radiation or radio (including Bluetooth), although many such cordless interfaces are themselves connected through the aforementioned wired serial buses.

While the electrical interface and the format of the data transmitted by commonly available mice is currently standardized on USB, in the past it varied between different manufacturers. A bus mouse used a dedicated interface card for connection to an IBM PC or compatible computer.

Mouse use in DOS applications became more common after the introduction of the Microsoft Mouse, largely because Microsoft provided an open standard for communication between applications and mouse driver software. Thus, any application written to use the Microsoft standard could use a mouse with a driver that implements the same API, even if the mouse hardware itself was incompatible with Microsoft's. This driver provides the state of the buttons and the distance the mouse has moved in units that its documentation calls "mickeys".

Early Mice

- In the 1970s, the Xerox Alto mouse, and in the 1980s the Xerox optical mouse, used a quadrature-encoded X and Y interface. This two-bit encoding per dimension had the property that only one bit of the two would change at a time, like a Gray code or Johnson counter, so that the transitions would not be misinterpreted when asynchronously sampled.
- The earliest mass-market mice, such as on the original Macintosh, Amiga, and Atari ST mice used a D-subminiature 9-pin connector to send the quadrature-encoded X and Y axis signals directly, plus one pin per mouse button. The mouse was a simple optomechanical device, and the decoding circuitry was all in the main computer.
- The DE-9 connectors were designed to be electrically compatible with the joysticks popular on numerous 8-bit systems, such as the Commodore 64 and the Atari 2600. Although the ports could be used for both purposes, the signals must be interpreted differently. As a result, plugging a mouse into a joystick port causes the "joystick" to continuously move in some direction, even if the mouse stays still, whereas plugging a joystick into a mouse port causes the "mouse" to only be able to move a single pixel in each direction.

Serial Interface and Protocol

- Because the IBM PC did not have a quadrature decoder built in, early PC mice used the RS-232C serial port to communicate encoded mouse movements, as well as provide power to the mouse's circuits. The Mouse Systems Corporation (MSC) version used a five-byte protocol and supported three buttons. The Microsoft version used a three-byte protocol and supported two buttons. Due to the incompatibility between the two protocols, some manufacturers sold serial mice with a mode switch: "PC" for MSC mode, "MS" for Microsoft mode.

Apple Desktop Bus

- In 1986 Apple first implemented the Apple Desktop Bus allowing the daisy chaining of up to 16 devices, including mice and other devices on the same bus with no configuration whatsoever. Featuring only a single data pin, the bus used a purely polled approach to device communications and survived as the standard on mainstream models (including a number of non-Apple workstations) until 1998 when Apple's iMac line of computers joined the industry-wide switch to using USB. Beginning with the Bronze Keyboard PowerBook G3 in May 1999, Apple dropped the external ADB port in favor of USB, but retained an internal ADB connection in the PowerBook G4 for communication with its built-in keyboard and trackpad until early 2005.

USB

- Almost all wired mice today use USB and the USB human interface device class for communication.

Cordless or Wireless

- Cordless or wireless mice transmit data via radio. Some mice connect to the computer through Bluetooth or Wi-Fi, while others use a receiver that plugs into the computer, for example through a USB port.
- Many mice that use a USB receiver have a storage compartment for it inside the mouse. Some "nano receivers" are designed to be small enough to remain plugged into a laptop during transport, while still being large enough to easily remove.

Operating System Support

MS-DOS and Windows 1.0 support connecting a mouse such as a Microsoft Mouse via multiple interfaces: BallPoint, Bus (InPort), Serial port or PS/2. Windows 98 added built-in support for USB Human Interface Device class (USB HID), with native vertical scrolling support. Windows 2000 and Windows Me expanded this built-in support to 5-button mice.

Windows XP Service Pack 2 introduced a Bluetooth stack, allowing Bluetooth mice to be used without any USB receivers. Windows Vista added native support for horizontal scrolling and standardized wheel movement granularity for finer scrolling. Windows 8 introduced BLE (Bluetooth Low Energy) mouse/HID support.

Multiple-Mouse Systems

Some systems allow two or more mice to be used at once as input devices. Late-1980s era home computers such as the Amiga used this to allow computer games with two players interacting on the same computer (Lemmings and The Settlers for example). The same idea is sometimes used in collaborative software, e.g. to simulate a whiteboard that multiple users can draw on without passing a single mouse around.

Microsoft Windows, since Windows 98, has supported multiple simultaneous pointing devices. Because Windows only provides a single screen cursor, using more than one device at the same time requires cooperation of users or applications designed for multiple input devices.

Multiple mice are often used in multi-user gaming in addition to specially designed devices that provide several input interfaces. Windows also has full support for multiple input/mouse configurations for multi-user environments. Starting with Windows XP, Microsoft introduced an SDK for developing applications that allow multiple input devices to be used at the same time with independent cursors and independent input points. However, it no longer appears to be available.

The introduction of Windows Vista and Microsoft Surface (now known as Microsoft PixelSense) introduced a new set of input APIs that were adopted into Windows 7, allowing for 50 points/cursors, all controlled by independent users. The new input points provide traditional mouse input; however, they were designed with other input technologies like touch and image in mind. They inherently offer 3D coordinates along with pressure, size, tilt, angle, mask, and even an image bitmap to see and recognize the input point/object on the screen.

As of 2009, Linux distributions and other operating systems that use X.Org, such as OpenSolaris and FreeBSD, support 255 cursors/input points through Multi-Pointer X. However, currently no window managers support Multi-Pointer X leaving it relegated to custom software usage.

There have also been propositions of having a single operator use two mice simultaneously as a more sophisticated means of controlling various graphics and multimedia applications.

Buttons

Mouse buttons are microswitches which can be pressed to select or interact with an element of a graphical user interface, producing a distinctive clicking sound. Since around the late 1990s, the three-button scrollmouse has become the de facto standard. Users most commonly employ the second button to invoke a contextual menu in the computer's software user interface, which contains options specifically tailored to the interface element over which the mouse cursor currently sits. By default, the primary mouse button sits located on the left-hand side of the mouse, for the benefit of right-handed users; left-handed users can usually reverse this configuration via software.

Scrolling

Nearly all mice now have an integrated input primarily intended for scrolling on top, usually a single-axis digital wheel or rocker switch which can also be depressed to act as a third button. Though less common, many mice instead have two-axis inputs such as a tiltable wheel, trackball, or touchpad. Those with a trackball may be designed to stay stationary, using the trackball instead of moving the mouse.

Speed

Mickeys per second is a unit of measurement for the speed and movement direction of a computer mouse, where direction is often expressed as "horizontal" versus "vertical" mickey count. However, speed can also refer to the ratio between how many pixels the cursor moves on the screen and how far the mouse moves on the mouse pad, which may be expressed as pixels per mickey, pixels per inch, or pixels per centimeter.

The computer industry often measures mouse sensitivity in terms of counts per inch (CPI), commonly expressed as dots per inch (DPI) – the number of steps the mouse will report when it moves one inch. In early mice, this specification was called pulses per inch (ppi). The mickey originally referred to one of these counts, or one resolvable step of motion. If the default mouse-tracking condition involves moving the cursor by one screen-pixel or dot on-screen per reported step, then the CPI does equate to DPI: dots of cursor motion per inch of mouse motion. The CPI or DPI as reported by manufacturers depends on how they make the mouse; the higher the CPI, the faster the cursor moves with mouse movement. However, software can adjust the mouse sensitivity, making the cursor move faster or slower than its CPI. As of 2007, software can change the speed of the cursor dynamically, taking into account the mouse's absolute speed and the movement from the last stop-point. In most software, an example being the Windows platforms, this setting is named "speed", referring to "cursor precision". However, some operating systems name this setting "acceleration", the typical Apple OS designation. This term is incorrect. Mouse acceleration in most mouse software refers to the change in speed of the cursor over time while the mouse movement is constant.

For simple software, when the mouse starts to move, the software will count the number of "counts" or "mickeys" received from the mouse and will move the cursor across the screen by that number of pixels (or multiplied by a rate factor, typically less than 1). The cursor will move slowly on the screen, with good precision. When the movement of the mouse passes the value set for some threshold, the software will start to move the cursor faster, with a greater rate factor. Usually, the user can set the value of the second rate factor by changing the "acceleration" setting.

Operating systems sometimes apply acceleration, referred to as "ballistics", to the motion reported by the mouse. For example, versions of Windows prior to Windows XP doubled reported values above a configurable threshold, and then optionally doubled them again above a second configurable threshold. These doublings applied separately in the X and Y directions, resulting in very nonlinear response.

Mousepads

Engelbart's original mouse did not require a mousepad; the mouse had two large wheels which could roll on virtually any surface. However, most subsequent mechanical mice starting with the steel roller ball mouse have required a mousepad for optimal performance.

The mousepad, the most common mouse accessory, appears most commonly in conjunction with mechanical mice, because to roll smoothly the ball requires more friction than common desk surfaces usually provide. So-called "hard mousepads" for gamers or optical/laser mice also exist.

Most optical and laser mice do not require a pad, the notable exception being early optical mice which relied on a grid on the pad to detect movement (e.g., Mouse Systems). Whether to use a hard or soft mousepad with an optical mouse is largely a matter of personal preference. One exception occurs when the desk surface creates problems for the optical or laser tracking, for example, a transparent or reflective surface, such as glass.

Some mice also come with small "pads" attached to the bottom surface, also called mouse feet or mouse skates, that help the user slide the mouse smoothly across surfaces.

Computer Keyboard

A computer keyboard is a peripheral input device modeled after the typewriter keyboard which uses an arrangement of buttons or keys to act as mechanical levers or electronic switches. Replacing early punched cards and paper tape technology, interaction via teleprinter-style keyboards have been the main input method for computers since the 1970s, supplemented by the computer mouse since the 1980s.

Keyboard keys (buttons) typically have a set of characters engraved or printed on them, and each press of a key typically corresponds to a single written symbol. However, producing some symbols may require pressing and holding several keys simultaneously or in sequence. While most keys produce characters (letters, numbers or symbols), other keys (such as the escape key) can prompt the computer to execute system commands. In a modern computer, the interpretation of key presses is generally left to the software: the information sent to the computer, the scan code, tells it only which physical key (or keys) was pressed or released.

In normal usage, the keyboard is used as a text entry interface for typing text, numbers, and symbols into application software such as a word processor, web browser or social media app. Touchscreens use virtual keyboards.

History

Typewriters are the definitive ancestor of all key-based text entry devices, but the computer keyboard as a device for electromechanical data entry and communication largely comes from the utility of two devices: teleprinters (or teletypes) and keypunches. It was through such devices that modern computer keyboards inherited their layouts.

As early as the 1870s, teleprinter-like devices were used to simultaneously type and transmit stock market text data from the keyboard across telegraph lines to stock ticker machines to be immediately copied and displayed onto ticker tape. The teleprinter, in its more contemporary form, was developed from 1907 to 1910 by American mechanical engineer Charles Krum and his son Howard, with early contributions by electrical engineer Frank Pearne. Earlier models were developed separately by individuals such as Royal Earl House and Frederick G. Creed.

Earlier, Herman Hollerith developed the first keypunch devices, which soon evolved to include keys for text and number entry akin to normal typewriters by the 1930s.

The keyboard on the teleprinter played a strong role in point-to-point and point-to-multipoint communication for most of the 20th century, while the keyboard on the keypunch device played a strong role in data entry and storage for just as long.

The development of some of the earliest computers incorporated electric typewriter keyboards: the development of the ENIAC computer incorporated a keypunch device as both the input and paper-based output device, and the BINAC computer made use of an electromechanically controlled typewriter for both data entry onto magnetic tape (instead of paper) and data output.

The keyboard remained the primary, most integrated computer peripheral well into the era of personal computing until the introduction of the mouse as a consumer device in 1984. By this time, text-only user interfaces with sparse graphics gave way to comparatively graphics-rich icons on screen.

However, keyboards remain central to human-computer interaction to the present though mobile personal computing devices such as smartphones and tablets use a virtual keyboard.

Types and Standards

Different types of keyboards are available and each is designed with a focus on specific features that suit particular needs. Today, most full-size keyboards use one of three different mechanical layouts, usually referred to as simply *ISO* (ISO/IEC 9995-2), *ANSI* (ANSI-INCITS 154-1988), and *JIS* (JIS X 6002-1980), referring roughly to the organizations issuing the relevant worldwide, United States, and Japanese standards, respectively. (In fact, the mechanical layouts referred such as "ISO" and "ANSI" comply to the primary recommendations in the named standards, while each of these standards in fact also allows the other way.) ANSI standard alphanumeric keyboards have keys that are on three-quarter inch centers (0.75 inches (19 mm)), and have a key travel of at least 0.15 inches (3.8 mm).

Modern keyboard models contain a set number of total keys according to their given standard, described as 101, 104, 105, etc. and sold as "Full-size" keyboards. Modern keyboards matching US conventions typically have 104 keys while the 105 key layout is the norm in the rest of the world. This number is not always followed, and individual keys or whole sections are commonly skipped for the sake of compactness or user preference.

The most common choice is to not include the numpad, which can usually be fully replaced by the alphanumeric section; such designs are referred to as "tenkeyless" (or TKL). Laptops and wireless peripherals often lack duplicate keys and ones seldom used. Function- and arrow keys are nearly always present.

Another factor determining the size of a keyboard is the size and spacing of the keys. The reduction is limited by the practical consideration that the keys must be large enough to be easily pressed by fingers. Alternatively, a tool is used for pressing small keys.

- **Desktop or Full-Size:** Desktop computer keyboards include alphabetic characters and numerals (and usually additionally a numeric keypad), typographical symbols and punctuation marks, one or more currency symbols and other special characters, diacritics and a variety of function keys. The repertoire of glyphs engraved on the keys of a keyboard accords with national conventions and language needs. Computer keyboards are similar to electric-typewriter keyboards but contain additional keys, such as the command key or Windows keys.
- **Laptop-Size:** Keyboards on laptops and notebook computers usually have a shorter travel distance for the keystroke, shorter over travel distance, and a reduced set of keys. They may not have a numeric keypad, and the function keys may be placed in locations that differ from their placement on a standard, full-sized keyboard. The switch mechanism for a laptop keyboard is more likely to be a scissor switch than a rubber dome; this is opposite the trend for full-size keyboards.
- **Flexible Keyboards:** Flexible keyboards are a junction between normal type and laptop type keyboards: normal from the full arrangement of keys, and laptop from the short key distance. Additionally, the flexibility allows the user to fold/roll the keyboard for better storage and transfer. However, for typing the keyboard must be resting on a hard surface. The vast majority of flexible keyboards in the market are made from silicone; this material makes them water- and dust-proof.

This is useful in hospitals, where keyboards are subjected to frequent washing, and other dirty or must-be-clean environments.

- **Handheld:** Handheld ergonomic keyboards are designed to be held like a game controller, and can be used as such, instead of laid out flat on top of a table surface. Typically handheld keyboards hold all the alphanumeric keys and symbols that a standard keyboard would have, yet only be accessed by pressing two sets of keys at once; one acting as a function key similar to a 'Shift' key that would allow for capital letters on a standard keyboard. Handheld keyboards allow the user the ability to move around a room or to lean back on a chair while also being able to type in front or away from the computer. Some variations of handheld ergonomic keyboards also include a trackball mouse that allow mouse movement and typing included in one handheld device.
- **Thumb-Sized:** Smaller external keyboards have been introduced for devices without a built-in keyboard, such as PDAs, and smartphones. Small keyboards are also useful where there is a limited workspace. A thumb keyboard (thumb board) is used in some personal digital assistants such as the Palm Treo and BlackBerry and some Ultra-Mobile PCs such as the OQO. Numeric keyboards contain only numbers, mathematical symbols for addition, subtraction, multiplication, and division, a decimal point, and several function keys. They are often used to facilitate data entry with smaller keyboards that do not have a numeric keypad, commonly those of laptop computers. These keys are collectively known as a numeric pad, numeric keys, or a numeric keypad, and it can consist of the following types of keys: Arithmetic operators, numbers, arrow keys, Navigation keys, Num Lock and Enter key.
- **Multifunctional:** Multifunctional keyboards provide additional function beyond the standard keyboard. Many are programmable, configurable computer keyboards and some control multiple PCs, workstations and other information sources, usually in multi-screen work environments. Users have additional key functions as well as the standard functions and can typically use a single keyboard and mouse to access multiple sources. Multifunctional keyboards may feature customised keypads, fully programmable function or soft keys for macros/pre-sets, biometric or smart card readers, trackballs, etc. New generation multifunctional keyboards feature a touchscreen display to stream video, control audio visual media and alarms, execute application inputs, configure individual desktop environments, etc.

Multifunctional keyboards may also permit users to share access to PCs and other information sources. Multiple interfaces (serial, USB, audio, Ethernet, etc.) are used to integrate external devices. Some multifunctional keyboards are also used to directly and intuitively control video walls. Common environments for multifunctional keyboards are complex, high-performance workplaces for financial traders and control room operators (emergency services, security, air traffic management; industry, utilities management, etc.).

Non-Standard Layout and Special-Use Types

- **One-Handed Keyboards:** Many keyboards have been designed for one-handed operation. The first one, a chorded keyboard, was invented by Douglas Engelbart. Other types of one-handed keyboards include the FrogPad, the Half-keyboard, and one-handed Dvorak keyboard layouts designed for one hand typing. While other keyboards generally associate one action with each key, chorded keyboards associate actions with combinations of key presses. Since there are many combinations available, chorded keyboards can effectively produce more actions on a board with fewer keys. Court reporters' stenotype machines use chorded keyboards to enable them to enter text much faster by typing a syllable with each stroke instead of one letter at a time. The fastest typists (as of 2007) use a stenograph, a kind of chorded keyboard used by most court reporters and closed-caption reporters. Some chorded keyboards are also made for use in situations where fewer keys are preferable, such as on devices that can be used with only one hand, and on small mobile devices that don't have room for larger keyboards. Chorded keyboards are less desirable in many cases because it usually takes practice and memorization of the combinations to become proficient.
- **Virtual:** Virtual keyboards, sometimes called on-screen keyboards (rarely software keyboards), consist of computer programs that display an image of a keyboard on the screen. Another input device such as a mouse or a touchscreen can be used to operate each virtual key to enter text. Virtual keyboards have become very popular in touchscreen enabled cell phones due to the additional cost and space requirements of other types of hardware keyboards. Microsoft Windows, Mac OS X, and some varieties of Linux include on-screen keyboards that can be controlled with the mouse. In these, the mouse has to be maneuvered onto the on-screen letters given by the software. On the click of a letter, the software writes the respective letter in the respective spot.
- **Projection:** Projection keyboards project an image of keys, usually with

a laser, onto a flat surface. The device then uses a camera or infrared sensor to "watch" where the user's fingers move, and will count a key as being pressed when it "sees" the user's finger touch the projected image. Projection keyboards can simulate a full size keyboard from a very small projector. Because the "keys" are simply projected images, they cannot be felt when pressed. Users of projected keyboards often experience increased discomfort in their fingertips because of the lack of "give" when typing. A flat, non-reflective surface is also required for the keys to be projected. Most projection keyboards are made for use with PDAs and smartphones due to their small form factor.

- **Optical Keyboard Technology:** Also known as photo-optical keyboard, light responsive keyboard, photo-electric keyboard and optical key actuation detection technology. An optical keyboard technology utilizes LEDs and photo sensors to optically detect actuated keys. Most commonly the emitters and sensors are located in the perimeter, mounted on a small PCB. The light is directed from side to side of the keyboard interior and it can only be blocked by the actuated keys. Most optical keyboards require at least 2 beams (most commonly vertical beam and horizontal beam) to determine the actuated key. Some optical keyboards use a special key structure that blocks the light in a certain pattern, allowing only one beam per row of keys (most commonly horizontal beam).

1.1.4 The Rise of the Internet and Mobile Computing

Web Interfaces

The late 1990s and early 2000s brought the explosion of the internet, fundamentally changing how humans interacted with machines. Web browsers became the primary interface for accessing information and services, and websites evolved from static pages to dynamic, interactive platforms.

The internet facilitated a shift from isolated, single-user interactions to interconnected, multi-user experiences.

User Interface

In the industrial design field of human–computer interaction, a user interface (UI) is the space where interactions between humans and machines occur. The goal of this interaction is to allow effective operation and control of the machine from the human end, while the machine simultaneously feeds back information that aids the operators' decision-making process. Examples of this broad concept of user interfaces include the interactive aspects of computer operating systems, hand tools, heavy machinery operator controls and process controls. The design considerations applicable when creating user interfaces are related to, or involve such disciplines as, ergonomics and psychology.

Generally, the goal of user interface design is to produce a user interface that makes it easy, efficient, and enjoyable (user-friendly) to operate a machine in the way which produces the desired result (i.e., maximum usability). This generally means that the operator needs to provide minimal input to achieve the desired output, and also that the machine minimizes undesired outputs to the user.

User interfaces are composed of one or more layers, including a human-machine interface (HMI) that typically interfaces machines with physical input hardware (such as keyboards, mice, or game pads) and output hardware (such as computer monitors, speakers, and printers). A device that implements an HMI is called a human interface device (HID). User interfaces that dispense with the physical movement of body parts as an intermediary step between the brain and the machine use no input or output devices except electrodes alone; they are called brain–computer interfaces (BCIs) or brain–machine interfaces (BMIs).

Other terms for human–machine interfaces are man–machine interface (MMI) and, when the machine in question is a computer, human–computer interface. Additional UI layers may interact with one or more human senses, including: tactile UI (touch), visual UI (sight), auditory UI (sound), olfactory UI (smell), equilibria UI (balance), and gustatory UI (taste).

Composite user interfaces (CUIs) are UIs that interact with two or more senses. The most common CUI is a *graphical user interface* (GUI), which is composed of a tactile UI and a visual UI capable of displaying graphics. When sound is added to a GUI, it becomes a *multimedia user interface* (MUI). There are three broad categories of CUI: *standard*, *virtual* and *augmented*. Standard CUI use standard human interface devices like keyboards, mice, and computer monitors. When the CUI blocks out the real world to create a virtual reality, the CUI is virtual and uses a *virtual reality interface*. When the CUI does not block out the real world and creates augmented reality, the CUI is augmented and uses an *augmented reality interface*. When a UI interacts with all human senses, it is called a qualia interface, named after the theory of qualia.

CUI may also be classified by how many senses they interact with as either an X-sense virtual reality interface or X-sense augmented reality interface, where X is the number of senses interfaced with. For example, a Smell-O-Vision is a 3-sense (3S) Standard CUI with visual display, sound and smells; when *virtual reality interfaces* interface with smells and touch it is said to be a 4-sense (4S) virtual reality interface; and when *augmented reality interfaces* interface with smells and touch it is said to be a 4-sense (4S) augmented reality interface.

The user interface or *human-machine interface* is the part of the machine that handles the human-machine interaction. Membrane switches, rubber keypads and touchscreens are examples of the physical part of the Human Machine Interface which we can see and touch.

In complex systems, the human-machine interface is typically computerized. The term *human-computer interface* refers to this kind of system. In the context of computing, the term typically extends as well to the software dedicated to control the physical elements used for human-computer interaction.

The engineering of human-machine interfaces is enhanced by considering ergonomics (human factors). The corresponding disciplines are human factors engineering (HFE) and usability engineering (UE) which is part of systems engineering.

Tools used for incorporating human factors in the interface design are developed based on knowledge of computer science, such as computer graphics, operating systems, programming languages. Nowadays, we use the expression graphical user interface for human-machine interface on computers, as nearly all of them are now using graphics. Multimodal interfaces allow users to interact using more than one modality of user input.

Terminology

There is a difference between a user interface and an operator interface or a human-machine interface (HMI).

- The term "user interface" is often used in the context of (personal) computer systems and electronic devices.
  - Where a network of equipment or computers are interlinked through an MES (Manufacturing Execution System)-or Host to display information.
  - A human-machine interface (HMI) is typically local to one machine or piece of equipment, and is the interface method between the human and the equipment/machine. An operator interface is the interface method by which multiple pieces of equipment, linked by a host control system, are accessed or controlled.
  - The system may expose several user interfaces to serve different kinds of users. For example, a computerized library database might provide two user interfaces, one for library patrons (limited set of functions, optimized for ease of use) and the other for library personnel (wide set of functions, optimized for efficiency).
- The user interface of a mechanical system, a vehicle or an industrial installation is sometimes referred to as the human-machine interface (HMI). HMI is a modification of the original term MMI (man-machine

> interface). In practice, the abbreviation MMI is still frequently used although some may claim that MMI stands for something different now. Another abbreviation is HCI, but is more commonly used for human-computer interaction. Other terms used are operator interface console (OIC) and operator interface terminal (OIT). However it is abbreviated, the terms refer to the 'layer' that separates a human that is operating a machine from the machine itself. Without a clean and usable interface, humans would not be able to interact with information systems.

In science fiction, HMI is sometimes used to refer to what is better described as a direct neural interface. However, this latter usage is seeing increasing application in the real-life use of (medical) prostheses—the artificial extension that replaces a missing body part (e.g., cochlear implants).

In some circumstances, computers might observe the user and react according to their actions without specific commands. A means of tracking parts of the body is required, and sensors noting the position of the head, direction of gaze and so on have been used experimentally. This is particularly relevant to immersive interfaces.

1945–1968: Batch Interface

In the batch era, computing power was extremely scarce and expensive. User interfaces were rudimentary. Users had to accommodate computers rather than the other way around; user interfaces were considered overhead, and software was designed to keep the processor at maximum utilization with as little overhead as possible.

The input side of the user interfaces for batch machines was mainly punched cards or equivalent media like paper tape. The output side added line printers to these media. With the limited exception of the system operator's console, human beings did not interact with batch machines in real time at all.

Submitting a job to a batch machine involved, first, preparing a deck of punched cards describing a program and a dataset. Punching the program cards was not done on the computer itself, but on keypunches, specialized typewriter-like machines that were notoriously bulky, unforgiving, and prone to mechanical failure. The software interface was similarly unforgiving, with very strict syntaxes meant to be parsed by the smallest possible compilers and interpreters.

Once the cards were punched, one would drop them in a job queue and wait. Eventually, operators would feed the deck to the computer, perhaps mounting magnetic tapes to supply another dataset or helper software. The job would generate a printout, containing final results or an abort notice with an attached error log. Successful runs might also write a result on magnetic tape or generate some data cards to be used in a later computation.

The turnaround time for a single job often spanned entire days. If one was very lucky, it might be hours; there was no real-time response. But there were worse fates than the card queue; some computers required an even more tedious and error-prone process of toggling in programs in binary code using console switches. The very earliest machines had to be partly rewired to incorporate program logic into themselves, using devices known as plugboards.

Early batch systems gave the currently running job the entire computer; program decks and tapes had to include what we would now think of as operating system code to talk to I/O devices and do whatever other housekeeping was needed. Midway through the batch period, after 1957, various groups began to experiment with so-called "load-and-go" systems. These used a monitor program which was always resident on the computer. Programs could call the monitor for services. Another function of the monitor was to do better error checking on submitted jobs, catching errors earlier and more intelligently and generating more useful feedback to the users. Thus, monitors represented the first step towards both operating systems and explicitly designed user interfaces.

1969–present: Command-Line User Interface

Command-line interfaces (CLIs) evolved from batch monitors connected to the system console. Their interaction model was a series of request-response transactions, with requests expressed as textual commands in a specialized vocabulary. Latency was far lower than for batch systems, dropping from days or hours to seconds. Accordingly, command-line systems allowed the user to change their mind about later stages of the transaction in response to real-time or near-real-time feedback on earlier results. Software could be exploratory and interactive in ways not possible before. But these interfaces still placed a relatively heavy mnemonic load on the user, requiring a serious investment of effort and learning time to master.

The earliest command-line systems combined teleprinters with computers, adapting a mature technology that had proven effective for mediating the transfer of information over wires between human beings. Teleprinters had originally been invented as devices for automatic telegraph transmission and reception; they had a history going back to 1902 and had already become well-established in newsrooms and elsewhere by 1920. In reusing them, economy was certainly a consideration, but psychology and the rule of least surprise mattered as well; teleprinters provided a point of interface with the system that was familiar to many engineers and users.

The widespread adoption of video-display terminals (VDTs) in the mid-1970s ushered in the second phase of command-line systems. These cut latency further, because characters could be thrown on the phosphor dots of a screen more quickly than a printer head or carriage can move. They helped quell conservative resistance to interactive programming by cutting ink and paper consumables out of the cost picture, and were to the first TV generation of the late 1950s and 60s even more iconic and comfortable than teleprinters had been to the computer pioneers of the 1940s.

Just as importantly, the existence of an accessible screen—a two-dimensional display of text that could be rapidly and reversibly modified—made it economical for software designers to deploy interfaces that could be described as visual rather than textual.

The pioneering applications of this kind were computer games and text editors; close descendants of some of the earliest specimens, such as rogue(6), and vi(1), are still a live part of Unix tradition.

1985: SAA User Interface or Text-Based User Interface

In 1985, with the beginning of Microsoft Windows and other graphical user interfaces, IBM created what is called the Systems Application Architecture (SAA) standard which include the Common User Access (CUA) derivative. CUA successfully created what we know and use today in Windows, and most of the more recent DOS or Windows Console Applications will use that standard as well.

This defined that a pulldown menu system should be at the top of the screen, status bar at the bottom, shortcut keys should stay the same for all common functionality (F2 to Open for example would work in all applications that followed the SAA standard). This greatly helped the speed at which users could learn an application so it caught on quick and became an industry standard.

1968–Present: Graphical User Interface

- **1968:** Douglas Engelbart demonstrated NLS, a system which uses a mouse, pointers, hypertext, and multiple windows.
- **1970:** Researchers at Xerox Palo Alto Research Center (many from SRI) develop WIMP paradigm (Windows, Icons, Menus, Pointers)
- **1973:** Xerox Alto: commercial failure due to expense, poor user interface, and lack of programs
- **1979:** Steve Jobs and other Apple engineers visit Xerox PARC. Though Pirates of Silicon Valley dramatizes the events, Apple had already been working on developing a GUI, such as the Macintosh and Lisa projects, before the visit.
- **1981:** Xerox Star: focus on WYSIWYG. Commercial failure (25K sold) due to cost ($16K each), performance (minutes to save a file, couple of

hours to recover from crash), and poor marketing

- **1982:** Rob Pike and others at Bell Labs designed Blit, which was released in 1984 by AT&T and Teletype as DMD 5620 terminal.
- **1984:** Apple Macintosh popularizes the GUI. Super Bowl commercial shown twice, was the most expensive commercial ever made at that time
- **1984:** MIT's X Window System: hardware-independent platform and networking protocol for developing GUIs on UNIX-like systems
- **1985:** Windows 1.0 – provided GUI interface to MS-DOS. No overlapping windows (tiled instead).
- **1985:** Microsoft and IBM start work on OS/2 meant to eventually replace MS-DOS and Windows
- **1986:** Apple threatens to sue Digital Research because their GUI desktop looked too much like Apple's Mac.
- **1987:** Windows 2.0 – Overlapping and resizable windows, keyboard and mouse enhancements
- **1987:** Macintosh II: first full-color Mac
- **1988:** OS/2 1.10 Standard Edition (SE) has GUI written by Microsoft, looks a lot like Windows 2

Interface Design

Primary methods used in the interface design include prototyping and simulation. Typical human–machine interface design consists of the following stages: interaction specification, interface software specification and prototyping:

- Common practices for interaction specification include user-centered design, persona, activity-oriented design, scenario-based design, and resiliency design.
- Common practices for interface software specification include use cases and constrain enforcement by interaction protocols (intended to avoid use errors).
- Common practices for prototyping are based on libraries of interface elements (controls, decoration, etc.).

In broad terms, interfaces generally regarded as user friendly, efficient, intuitive, etc. are typified by one or more particular qualities.

For the purpose of example, a non-exhaustive list of such characteristics follows:

- **Clarity:** The interface avoids ambiguity by making everything clear through language, flow, hierarchy and metaphors for visual elements.

- **Concision:** However ironically, the over-clarification of information—for instance, by labelling the majority, if not the entirety, of items displayed on-screen at once, and regardless of whether or not the user would in fact require a visual indicator of some kind in order to identify a given item—can, and, under most normal circumstances, most likely will lead to the obfuscation of whatever information.
- **Familiarity:** Even if someone uses an interface for the first time, certain elements can still be familiar. Real-life metaphors can be used to communicate meaning.
- **Responsiveness:** A good interface should not feel sluggish. This means that the interface should provide good feedback to the user about what's happening and whether the user's input is being successfully processed.
- **Consistency:** Keeping your interface consistent across your application is important because it allows users to recognize usage patterns.
- **Aesthetics:** While you do not need to make an interface attractive for it to do its job, making something look good will make the time your users spend using your application more enjoyable; and happier users can only be a good thing.
- **Efficiency:** Time is money, and a great interface should make the user more productive through shortcuts and good design.
- **Forgiveness:** A good interface should not punish users for their mistakes but should instead provide the means to remedy them.

The principle of least astonishment (POLA) is a general principle in the design of all kinds of interfaces. It is based on the idea that human beings can only pay full attention to one thing at one time, leading to the conclusion that novelty should be minimized. If an interface is used persistently, the user will unavoidably develop habits for using the interface. The designer's role can thus be characterized as ensuring the user forms good habits. If the designer is experienced with other interfaces, they will similarly develop habits, and often make unconscious assumptions regarding how the user will interact with the interface.

Peter Morville of Google designed the User Experience Honeycomb framework in 2004 when leading operations in user interface design. The framework was created to guide user interface design. It would act as a guideline for many web development students for a decade.

- **Usable:** Is the design of the system easy and simple to use? The application should feel familiar, and it should be easy to use.

- **Useful:** Does the application fulfill a need? A business's product or service needs to be useful.
- **Desirable:** Is the design of the application sleek and to the point? The aesthetics of the system should be attractive, and easy to translate.
- **Findable:** Are users able to quickly find the information they're looking for? Information needs to be findable and simple to navigate. A user should never have to hunt for your product or information.
- **Accessible:** Does the application support enlarged text without breaking the framework? An application should be accessible to those with disabilities.
- **Credible:** Does the application exhibit trustworthy security and company details? An application should be transparent, secure, and honest.
- **Valuable:** Does the end-user think it's valuable? If all 6 criteria are met, the end-user will find value and trust in the application.

Types

- **Attentive user interfaces** manage the user attention deciding when to interrupt the user, the kind of warnings, and the level of detail of the messages presented to the user.
- **Batch interfaces** are non-interactive user interfaces, where the user specifies all the details of the *batch job* in advance to batch processing, and receives the output when all the processing is done. The computer does not prompt for further input after the processing has started.
- **Command line interfaces** (CLIs) prompt the user to provide input by typing a command string with the computer keyboard and respond by outputting text to the computer monitor. Used by programmers and system administrators, in engineering and scientific environments, and by technically advanced personal computer users.
- **Conversational interfaces** enable users to command the computer with plain text English (e.g., via text messages, or chatbots) or voice commands, instead of graphic elements. These interfaces often emulate human-to-human conversations.
- **Conversational interface agents** attempt to personify the computer interface in the form of an animated person, robot, or other character (such as Microsoft's Clippy the paperclip), and present interactions in a

conversational form.

- **Crossing-based interfaces** are graphical user interfaces in which the primary task consists in crossing boundaries instead of pointing.
- **Direct manipulation interface** is the name of a general class of user interfaces that allow users to manipulate objects presented to them, using actions that correspond at least loosely to the physical world.
- **Gesture interfaces** are graphical user interfaces which accept input in a form of hand gestures, or mouse gestures sketched with a computer mouse or a stylus.
- **Graphical user interfaces** (GUI) accept input via devices such as a computer keyboard and mouse and provide articulated graphical output on the computer monitor. There are at least two different principles widely used in GUI design: Object-oriented user interfaces (OOUIs) and application-oriented interfaces.
- **Hardware interfaces** are the physical, spatial interfaces found on products in the real world from toasters, to car dashboards, to airplane cockpits. They are generally a mixture of knobs, buttons, sliders, switches, and touchscreens.
- **Holographic user interfaces** provide input to electronic or electro-mechanical devices by passing a finger through reproduced holographic images of what would otherwise be tactile controls of those devices, floating freely in the air, detected by a wave source and without tactile interaction.
- **Intelligent user interfaces** are human–machine interfaces that aim to improve the efficiency, effectiveness, and naturalness of human-machine interaction by representing, reasoning, and acting on models of the user, domain, task, discourse, and media (e.g., graphics, natural language, gesture).
- **Motion tracking interfaces** monitor the user's body motions and translate them into commands, currently being developed by Apple.
- **Multi-screen interfaces**, employ multiple displays to provide a more flexible interaction. This is often employed in computer game interaction in both the commercial arcades and more recently the handheld markets.
- **Natural-language interfaces** are used for search engines and on webpages. User types in a question and waits for a response.
- **Non-command user interfaces**, which observe the user to infer their needs and intentions, without requiring that they formulate explicit

commands.

- **Object-oriented user interfaces (OOUI)** are based on object-oriented programming metaphors, allowing users to manipulate simulated objects and their properties.
- **Permission-driven user interfaces** show or conceal menu options or functions depending on the user's level of permissions. The system is intended to improve the user experience by removing items that are unavailable to the user. A user who sees functions that are unavailable for use may become frustrated. It also provides an enhancement to security by hiding functional items from unauthorized persons.
- **Reflexive user interfaces** where the users control and redefine the entire system via the user interface alone, for instance to change its command verbs. Typically, this is only possible with very rich graphic user interfaces.
- **Search interface** is how the search box of a site is displayed, as well as the visual representation of the search results.
- **Tangible user interfaces**, which place a greater emphasis on touch and physical environment or its element.
- **Task-focused interfaces** are user interfaces which address the information overload problem of the desktop metaphor by making tasks, not files, the primary unit of interaction.
- **Text-based user interfaces** (TUIs) are user interfaces which interact via text. TUIs include command-line interfaces and text-based WIMP environments.
- **Touchscreens** are displays that accept input by touch of fingers or a stylus. Used in a growing amount of mobile devices and many types of point of sale, industrial processes and machines, self-service machines, etc.
- **Touch user interface** are graphical user interfaces using a touchpad or touchscreen display as a combined input and output device. They supplement or replace other forms of output with haptic feedback methods. Used in computerized simulators, etc.
- **Voice user interfaces**, which accept input and provide output by generating voice prompts. The user input is made by pressing keys or buttons, or responding verbally to the interface.
- **Web-based user interfaces or web user interfaces** (WUI) that accept input and provide output by generating web pages viewed by the user

using a web browser program.

- **Zero-input interfaces** get inputs from a set of sensors instead of querying the user with input dialogs.
- **Zooming user interfaces** are graphical user interfaces in which information objects are represented at different levels of scale and detail, and where the user can change the scale of the viewed area in order to show more detail.

Touchscreens and Mobile Devices

The introduction of smartphones and tablets in the late 2000s revolutionized HMI once again. Touchscreen technology, popularized by devices like the Apple iPhone, allowed for direct, multi-touch interactions. Users could now interact with their devices using gestures such as swiping, pinching, and tapping. This natural form of interaction made technology more accessible and integrated into daily life.

1.1.5 The Era of Natural User Interfaces (NUIs)

Voice Assistants

The 2010s saw the rise of voice-activated assistants like Apple's Siri, Amazon's Alexa, and Google Assistant. These technologies leveraged advances in natural language processing (NLP) and artificial intelligence (AI) to enable users to interact with machines using spoken language. Voice assistants marked a significant shift towards more natural and intuitive forms of interaction, allowing for hands-free control and accessibility.

Gesture and Motion Control

In parallel with voice technology, gesture and motion control interfaces emerged, enabling users to interact with machines through physical movements. Devices like the Microsoft Kinect and Leap Motion controller allowed for gesture-based interaction, expanding the possibilities for gaming, virtual reality (VR), and augmented reality (AR) applications.

These interfaces provided a more immersive and engaging experience, bridging the gap between the physical and digital worlds.

1.1.6 The Future of HMI: AI and Beyond

Artificial Intelligence and Machine Learning

As AI and machine learning technologies continue to advance, they are poised to further transform HMI. AI-powered interfaces can learn from user behavior, adapt to individual preferences, and anticipate needs, creating more personalized and efficient interactions. Chatbots, recommendation systems, and adaptive user interfaces are just a few examples of how AI is enhancing HMI.

Brain-Computer Interfaces (BCIs)

One of the most exciting frontiers in HMI is the development of brain-computer interfaces. BCIs aim to establish direct communication between the human brain and machines, bypassing traditional input methods. Companies like Neuralink are pioneering this technology, which has the potential to revolutionize fields such as healthcare, accessibility, and entertainment. While still in its infancy, BCI technology represents the ultimate goal of seamless human-machine integration.

Ethical and Societal Implications

The evolution of HMI also raises important ethical and societal questions. As machines become more integrated into our lives, issues of privacy, security, and digital divide come to the forefront.

Ensuring that HMI technologies are inclusive, ethical, and beneficial to all segments of society is a critical challenge for the future.

1.2 DEFINING HUMAN AUGMENTATION

Human augmentation encompasses a broad array of technologies and methodologies aimed at enhancing human abilities. This rapidly evolving field merges biology, technology, and engineering to amplify or restore human physical and cognitive capabilities. From medical implants to genetic engineering, human augmentation holds the promise of transformative changes in how we interact with the world and with each other.

Human augmentation represents a frontier in technology and medicine, promising to enhance human capabilities in unprecedented ways. From advanced prosthetics and cognitive enhancements to genetic engineering and AI integration, the possibilities are vast and transformative. However, these advancements come with significant ethical, societal, and regulatory challenges. As we move forward, it is essential to balance innovation with careful consideration of the broader implications, ensuring that human augmentation benefits society as a whole while safeguarding our fundamental values and identities.

1.2.1 Historical Context

Early Examples of Augmentation

The concept of enhancing human abilities is not new. Historically, humans have sought to improve their physical and cognitive functions through various means. Ancient prosthetics, like wooden toes found on Egyptian mummies, and the development of eyeglasses in the 13th century, are early examples of human augmentation. These innovations provided basic enhancements, compensating for lost or diminished abilities.

Industrial Revolution and Medical Advancements

The Industrial Revolution brought significant advancements in medical technology and prosthetics. The development of more sophisticated artificial limbs for injured soldiers during the World Wars marked a significant leap in augmentation technology. These early efforts laid the groundwork for contemporary prosthetics, which now integrate advanced materials and robotics to closely mimic natural limb function.

1.2.2 Modern Technologies in Human Augmentation

Prosthetics and Exoskeletons

Today's prosthetics are a far cry from their rudimentary predecessors. Modern prosthetics are equipped with advanced sensors and microprocessors, allowing for more natural movement and control. Some prosthetic limbs can even be controlled by neural signals, providing users with a seamless integration between their nervous system and the artificial limb.

Exoskeletons represent another significant advancement. These wearable devices can enhance physical strength and endurance, assisting individuals with mobility impairments and enabling workers to lift heavy loads with ease. Exoskeletons are finding applications in medical rehabilitation, military operations, and industrial settings.

Cognitive Enhancements

Cognitive augmentation technologies aim to enhance human mental capabilities. Nootropics, also known as smart drugs, are substances that can improve cognitive functions such as memory, creativity, and motivation. Advances in neurotechnology, including brain-computer interfaces (BCIs), offer more direct methods of cognitive enhancement.

BCIs can facilitate communication for individuals with severe disabilities and have the potential to boost cognitive functions in healthy individuals by directly interfacing with the brain's neural networks.

Sensory Augmentation

Sensory augmentation involves enhancing or restoring sensory capabilities. Cochlear implants are a well-known example, providing individuals with severe hearing loss the ability to perceive sound. Similarly, retinal implants are being developed to restore vision for those with certain types of blindness. Beyond restoring lost senses, there are experimental technologies aimed at extending human sensory perception, such as infrared vision or ultrasonic hearing, which could provide entirely new ways to interact with the environment.

1.2.3 Genetic and Biotechnological Augmentation

Gene Editing

Genome editing, or genome engineering, or gene editing, is a type of genetic engineering in which DNA is inserted, deleted, modified or replaced in the genome of a living organism. Unlike early genetic engineering techniques that randomly inserts genetic material into a host genome, genome editing targets the insertions to site-specific locations.

The basic mechanism involved in genetic manipulations through programmable nucleases is the recognition of target genomic loci and binding of effector DNA-binding domain (DBD), double-strand breaks (DSBs) in target DNA by the restriction endonucleases (FokI and Cas), and the repair of DSBs through homology-directed recombination (HDR) or non-homologous end joining (NHEJ).

History

Genome editing was pioneered in the 1990s, before the advent of the common current nuclease-based gene editing platforms but its use was limited by low efficiencies of editing. Genome editing with engineered nucleases, i.e. all three major classes of these enzymes—zinc finger nucleases (ZFNs), transcription activator-like effector nucleases (TALENs) and engineered meganucleases—were selected by *Nature Methods* as the 2011 Method of the Year. The CRISPR-Cas system was selected by *Science* as 2015 Breakthrough of the Year.

As of 2015 four families of engineered nucleases were used: meganucleases, zinc finger nucleases (ZFNs), transcription activator-like effector-based nucleases (TALEN), and the clustered regularly interspaced short palindromic repeats (CRISPR/Cas9) system. Nine genome editors were available as of 2017.

In 2018, the common methods for such editing used engineered nucleases, or "molecular scissors". These nucleases create site-specific double-strand breaks (DSBs) at desired locations in the genome.

The induced double-strand breaks are repaired through nonhomologous end-joining (NHEJ) or homologous recombination (HR), resulting in targeted mutations ('edits').

In May 2019, lawyers in China reported, in light of the purported creation by Chinese scientist He Jiankui of the first gene-edited humans, the drafting of regulations that anyone manipulating the human genome by gene-editing techniques, like CRISPR, would be held responsible for any related adverse consequences. A cautionary perspective on the possible blind spots and risks of CRISPR and related biotechnologies has been recently discussed, focusing on the stochastic nature of cellular control processes.

The University of Edinburgh Roslin Institute engineered pigs resistant to a virus that causes porcine reproductive and respiratory syndrome, which costs US and European pig farmers $2.6 billion annually.

In February 2020, a US trial safely showed CRISPR gene editing on 3 cancer patients. In 2020 Sicilian Rouge High GABA, a tomato that makes more of an amino acid said to promote relaxation, was approved for sale in Japan.

In 2021, England (not the rest of the UK) planned to remove restrictions on gene-edited plants and animals, moving from European Union-compliant regulation to rules closer to those of the US and some other countries. An April 2021 European Commission report found "strong indications" that the current regulatory regime was not appropriate for gene editing. Later in 2021, researchers announced a CRISPR alternative, labeled obligate mobile element–guided activity (OMEGA) proteins including IscB, IsrB and TnpB as endonucleases found in transposons, and guided by small ωRNAs.

Background

Genetic engineering as method of introducing new genetic elements into organisms has been around since the 1970s. One drawback of this technology has been the random nature with which the DNA is inserted into the hosts genome, which can impair or alter other genes within the organism. Although, several methods have been discovered which target the inserted genes to specific sites within an organism genome. It has also enabled the editing of specific sequences within a genome as well as reduced off target effects. This could be used for research purposes, by targeting mutations to specific genes, and in gene therapy. By inserting a functional gene into an organism and targeting it to replace the defective one it could be possible to cure certain genetic diseases.

Early methods to target genes to certain sites within a genome of an organism (called gene targeting) relied on homologous recombination (HR). By creating DNA constructs that contain a template that matches the targeted genome sequence it is possible that the HR processes within the cell will insert the construct at the desired location. Using this method on embryonic stem cells led to the development of transgenic mice with targeted genes knocked out. It has also been possible to knock in genes or alter gene expression patterns. In recognition of their discovery of how homologous recombination can be used to introduce genetic modifications in mice through embryonic stem cells, Mario Capecchi, Martin Evans and Oliver Smithies were awarded the 2007 Nobel Prize for Physiology or Medicine.

If a vital gene is knocked out it can prove lethal to the organism. In order to study the function of these genes site specific recombinases (SSR) were used. The two most common types are the Cre-LoxP and Flp-FRT systems. Cre recombinase is an enzyme that removes DNA by homologous recombination between binding sequences known as Lox-P sites. The Flip-FRT system operates in a similar way, with the Flip recombinase recognising FRT sequences. By crossing an organism containing the recombinase sites flanking the gene of interest with an organism that express the SSR under control of tissue specific promoters, it is possible to knock out or switch on genes only in certain cells. These techniques were also used to remove marker genes from transgenic animals. Further modifications of these systems allowed researchers to induce recombination only under certain conditions, allowing genes to be knocked out or expressed at desired times or stages of development.

Process

A common form of Genome editing relies on the concept of DNA double stranded break (DSB) repair mechanics. There are two major pathways that repair DSB; non-homologous end joining (NHEJ) and homology directed repair (HDR). NHEJ uses a variety of enzymes to directly join the DNA ends while the more accurate HDR uses a homologous sequence as a template for regeneration of missing DNA sequences at the break point. This can be exploited by creating a vector with the desired genetic elements within a sequence that is homologous to the flanking sequences of a DSB. This will result in the desired change being inserted at the site of the DSB. While HDR based gene editing is similar to the homologous recombination based gene targeting, the rate of recombination is increased by at least three orders of magnitude.

The key to genome editing is creating a DSB at a specific point within the genome. Commonly used restriction enzymes are effective at cutting DNA, but generally recognize and cut at multiple sites. To overcome this challenge and create site-specific DSB, three distinct classes of nucleases have been discovered and bioengineered to date. These are the Zinc finger nucleases (ZFNs), transcription-activator like effector nucleases (TALEN), meganucleases and the clustered regularly interspaced short palindromic repeats (CRISPR/Cas9) system.

Meganucleases, discovered in the late 1980s, are enzymes in the endonuclease family which are characterized by their capacity to recognize and cut large DNA sequences (from 14 to 40 base pairs). The most widespread and best known meganucleases are the proteins in the LAGLIDADG family, which owe their name to a conserved amino acid sequence.

Meganucleases, found commonly in microbial species, have the unique property of having very long recognition sequences (>14bp) thus making them naturally very specific. However, there is virtually no chance of finding the exact meganuclease required to act on a chosen specific DNA sequence. To overcome this challenge, mutagenesis and high throughput screening methods have been used to create meganuclease variants that recognize unique sequences. Others have been able to fuse various meganucleases and create hybrid enzymes that recognize a new sequence. Yet others have attempted to alter the DNA interacting aminoacids of the meganuclease to design sequence specific meganucelases in a method named rationally designed meganuclease. Another approach involves using computer models to try to predict as accurately as possible the activity of the modified meganucleases and the specificity of the recognized nucleic sequence.

A large bank containing several tens of thousands of protein units has been created. These units can be combined to obtain chimeric meganucleases that recognize the target site, thereby providing research and development tools that meet a wide range of needs (fundamental research, health, agriculture, industry, energy, etc.) These include the industrial-scale production of two meganucleases able to cleave the human XPC gene; mutations in this gene result in *Xeroderma pigmentosum*, a severe monogenic disorder that predisposes the patients to skin cancer and burns whenever their skin is exposed to UV rays.

Meganucleases have the benefit of causing less toxicity in cells than methods such as Zinc finger nuclease (ZFN), likely because of more stringent DNA sequence recognition; however, the construction of sequence-specific enzymes for all possible sequences is costly and time-consuming, as one is not benefiting from combinatorial possibilities that methods such as ZFNs and TALEN-based fusions utilize.

As opposed to meganucleases, the concept behind ZFNs and TALEN technology is based on a non-specific DNA cutting catalytic domain, which can then be linked to specific DNA sequence recognizing peptides such as zinc fingers and transcription activator-like effectors (TALEs). The first step to this was to find an endonuclease whose DNA recognition site and cleaving site were separate from each other, a situation that is not the most common among restriction enzymes. Once this enzyme was found, its cleaving portion could be separated which would be very non-specific as it would have no recognition ability. This portion could then be linked to sequence recognizing peptides that could lead to very high specificity.

Zinc finger motifs occur in several transcription factors. The zinc ion, found in 8% of all human proteins, plays an important role in the organization of their three-dimensional structure. In transcription factors, it is most often located at the protein-DNA interaction sites, where it stabilizes the motif. The C-terminal part of each finger is responsible for the specific recognition of the DNA sequence.

The recognized sequences are short, made up of around 3 base pairs, but by combining 6 to 8 zinc fingers whose recognition sites have been characterized, it is possible to obtain specific proteins for sequences of around 20 base pairs. It is therefore possible to control the expression of a specific gene. It has been demonstrated that this strategy can be used to promote a process of angiogenesis in animals. It is also possible to fuse a protein constructed in this way with the catalytic domain of an endonuclease in order to induce a targeted DNA break, and therefore to use these proteins as genome engineering tools.

The method generally adopted for this involves associating two DNA binding proteins – each containing 3 to 6 specifically chosen zinc fingers – with the catalytic domain of the FokI endonuclease which need to dimerize to cleave the double-strand DNA. The two proteins recognize two DNA sequences that are a few nucleotides apart. Linking the two zinc finger proteins to their respective sequences brings the two FokI domains closer together. FokI requires dimerization to have nuclease activity and this means the specificity increases dramatically as each nuclease partner would recognize a unique DNA sequence. To enhance this effect, FokI nucleases have been engineered that can only function as heterodimers.

Several approaches are used to design specific zinc finger nucleases for the chosen sequences. The most widespread involves combining zinc-finger units with known specificities (modular assembly). Various selection techniques, using bacteria, yeast or mammal cells have been developed to identify the combinations that offer the best specificity and the best cell tolerance. Although the direct genome-wide characterization of zinc finger nuclease activity has not been reported, an assay that measures the total number of double-strand DNA breaks in cells found that only one to two such breaks occur above background in cells treated with zinc finger nucleases with a 24 bp composite recognition site and obligate heterodimer FokI nuclease domains.

The heterodimer functioning nucleases would avoid the possibility of unwanted homodimer activity and thus increase specificity of the DSB. Although the nuclease portions of both ZFNs and TALEN constructs have similar properties, the difference between these engineered nucleases is in their DNA recognition peptide. ZFNs rely on Cys2-His2 zinc fingers and TALEN constructs on TALEs. Both of these DNA recognizing peptide domains have the characteristic that they are naturally found in combinations in their proteins. Cys2-His2 Zinc fingers typically happen in repeats that are 3 bp apart and are found in diverse combinations in a variety of nucleic acid interacting proteins such as transcription factors. Each finger of the Zinc finger domain is completely independent and the binding capacity of one finger is impacted by its neighbor. TALEs on the other hand are found in repeats with a one-to-one recognition ratio between the amino acids and the recognized nucleotide pairs. Because both zinc fingers and TALEs happen in repeated patterns, different combinations can be tried to create a wide variety of sequence specificities. Zinc fingers have been more established in these terms and approaches such as modular assembly (where Zinc fingers correlated with a triplet sequence are attached in a row to cover the required sequence), OPEN (low-stringency selection of peptide domains vs. triplet nucleotides followed by high-stringency selections of peptide combination vs. the final target in bacterial systems), and bacterial one-hybrid screening of zinc finger libraries among other methods have been used to make site specific nucleases.

Zinc finger nucleases are research and development tools that have already been used to modify a range of genomes, in particular by the laboratories in the Zinc Finger Consortium. The US company Sangamo BioSciences uses zinc finger nucleases to carry out research into the genetic engineering of stem cells and the modification of immune cells for therapeutic purposes. Modified T lymphocytes are currently undergoing phase I clinical trials to treat a type of brain tumor (glioblastoma) and in the fight against AIDS.

Transcription activator-like effector nucleases (TALENs) are specific DNA-binding proteins that feature an array of 33 or 34-amino acid repeats. TALENs are artificial restriction enzymes designed by fusing the DNA cutting domain of a nuclease to TALE domains, which can be tailored to specifically recognize a unique DNA sequence. These fusion proteins serve as readily targetable "DNA scissors" for gene editing applications that enable to perform targeted genome modifications such as sequence insertion, deletion, repair and replacement in living cells. The DNA binding domains, which can be designed to bind any desired DNA sequence, comes from TAL effectors, DNA-binding proteins excreted by plant pathogenic *Xanthomanos app.*

TAL effectors consists of repeated domains, each of which contains a highly conserved sequence of 34 amino acids, and recognize a single DNA nucleotide within the target site. The nuclease can create double strand breaks at the target site that can be repaired by error-prone non-homologous end-joining (NHEJ), resulting in gene disruptions through the introduction of small insertions or deletions. Each repeat is conserved, with the exception of the so-called repeat variable di-residues (RVDs) at amino acid positions 12 and 13. The RVDs determine the DNA sequence to which the TALE will bind.

This simple one-to-one correspondence between the TALE repeats and the corresponding DNA sequence makes the process of assembling repeat arrays to recognize novel DNA sequences straightforward. These TALEs can be fused to the catalytic domain from a DNA nuclease, FokI, to generate a transcription activator-like effector nuclease (TALEN). The resultant TALEN constructs combine specificity and activity, effectively generating engineered sequence-specific nucleases that bind and cleave DNA sequences only at pre-selected sites. The TALEN target recognition system is based on an easy-to-predict code. TAL nucleases are specific to their target due in part to the length of their 30+ base pairs binding site. TALEN can be performed within a 6 base pairs range of any single nucleotide in the entire genome.

TALEN constructs are used in a similar way to designed zinc finger nucleases, and have three advantages in targeted mutagenesis:

- DNA binding specificity is higher,
- off-target effects are lower, and
- construction of DNA-binding domains is easier.

CRISPRs (Clustered Regularly Interspaced Short Palindromic Repeats) are genetic elements that bacteria use as a kind of acquired immunity to protect against viruses.

They consist of short sequences that originate from viral genomes and have been incorporated into the bacterial genome. Cas (CRISPR associated proteins) process these sequences and cut matching viral DNA sequences. By introducing plasmids containing Cas genes and specifically constructed CRISPRs into eukaryotic cells, the eukaryotic genome can be cut at any desired position.

One of the earliest methods of efficiently editing nucleic acids employs nucleobase modifying enzymes directed by nucleic acid guide sequences was first described in the 1990s and has seen resurgence more recently.

This method has the advantage that it does not require breaking the genomic DNA strands, and thus avoids the random insertion and deletions associated with DNA strand breakage. It is only appropriate for precise editing requiring single nucleotide changes and has found to be highly efficient for this type of editing.

ARCUT stands for artificial restriction DNA cutter, it is a technique developed by Komiyama. This method uses pseudo-complementary peptide nucleic acid (pcPNA), for identifying cleavage site within the chromosome. Once pcPNA specifies the site, excision is carried out by cerium (CE) and EDTA (chemical mixture), which performs the splicing function.

Precision and Efficiency of Engineered Nucleases

Meganucleases method of gene editing is the least efficient of the methods mentioned above. Due to the nature of its DNA-binding element and the cleaving element, it is limited to recognizing one potential target every 1,000 nucleotides. ZFN was developed to overcome the limitations of meganuclease.

The number of possible targets ZFN can recognized was increased to one in every 140 nucleotides. However, both methods are unpredictable because of their DNA-binding elements affecting each other. As a result, high degrees of expertise and lengthy and costly validations processes are required.

TALE nucleases being the most precise and specific method yields a higher efficiency than the previous two methods. It achieves such efficiency because the DNA-binding element consists of an array of TALE subunits, each of them having the capability of recognizing a specific DNA nucleotide chain independent from others, resulting in a higher number of target sites with high precision. New TALE nucleases take about one week and a few hundred dollars to create, with specific expertise in molecular biology and protein engineering.

CRISPR nucleases have a slightly lower precision when compared to the TALE nucleases. This is caused by the need of having a specific nucleotide at one end in order to produce the guide RNA that CRISPR uses to repair the double-strand break it induces. It has been shown to be the quickest and cheapest method, only costing less than two hundred dollars and a few days of time. CRISPR also requires the least amount of expertise in molecular biology as the design lays in the guide RNA instead of the proteins. One major advantage that CRISPR has over the ZFN and TALEN methods is that it can be directed to target different DNA sequences using its ~80nt CRISPR sgRNAs, while both ZFN and TALEN methods required construction and testing of the proteins created for targeting each DNA sequence.

Because off-target activity of an active nuclease would have potentially dangerous consequences at the genetic and organismal levels, the precision of meganucleases, ZFNs, CRISPR, and TALEN-based fusions has been an active area of research. While variable figures have been reported, ZFNs tend to have more cytotoxicity than TALEN methods or RNA-guided nucleases, while TALEN and RNA-guided approaches tend to have the greatest efficiency and fewer off-target effects. Based on the maximum theoretical distance between DNA binding and nuclease activity, TALEN approaches result in the greatest precision.

Multiplex Automated Genomic Engineering (MAGE)

The methods for scientists and researchers wanting to study genomic diversity and all possible associated phenotypes were very slow, expensive, and inefficient. Prior to this new revolution, researchers would have to do single-gene manipulations and tweak the genome one little section at a time, observe the phenotype, and start the process over with a different single-gene manipulation. Therefore, researchers at the Wyss Institute at Harvard University designed the MAGE, a powerful technology that improves the process of in vivo genome editing. It allows for quick and efficient manipulations of a genome, all happening in a machine small enough to put on top of a small kitchen table. Those mutations combine with the variation that naturally occurs during cell mitosis creating billions of cellular mutations.

Chemically combined, synthetic single-stranded DNA (ssDNA) and a pool of oligionucleotides are introduced at targeted areas of the cell thereby creating genetic modifications. The cyclical process involves transformation of ssDNA (by electroporation) followed by outgrowth, during which bacteriophage homologous recombination proteins mediate annealing of ssDNAs to their genomic targets. Experiments targeting selective phenotypic markers are screened and identified by plating the cells on differential medias. Each cycle ultimately takes 2.5 hours to process, with additional time required to grow isogenic cultures and characterize mutations. By iteratively introducing libraries of mutagenic ssDNAs targeting multiple sites, MAGE can generate combinatorial genetic diversity in a cell population. There can be up to 50 genome edits, from single nucleotide base pairs to whole genome or gene networks simultaneously with results in a matter of days.

MAGE experiments can be divided into three classes, characterized by varying degrees of scale and complexity: (i) many target sites, single genetic mutations; (ii) single target site, many genetic mutations; and (iii) many target sites, many genetic mutations. An example of class three was reflected in 2009, where Church and colleagues were able to program *Escherichia coli* to produce five times the normal amount of lycopene, an antioxidant normally found in tomato seeds and linked to anti-cancer properties. They applied MAGE to optimize the 1-deoxy-D-xylulose 5-phosphate (DXP) metabolic pathway in *Escherichia coli* to overproduce isoprenoid lycopene. It took them about 3 days and just over $1,000 in materials. The ease, speed, and cost efficiency in which MAGE can alter genomes can transform how industries approach the manufacturing and production of important compounds in the bioengineering, bioenergy, biomedical engineering, synthetic biology, pharmaceutical, agricultural, and chemical industries.

Applications

As of 2012 efficient genome editing had been developed for a wide range of experimental systems ranging from plants to animals, often beyond clinical interest, and was becoming a standard experimental strategy in research labs. The recent generation of rat, zebrafish, maize and tobacco ZFN-mediated mutants and the improvements in TALEN-based approaches testify to the significance of the methods, and the list is expanding rapidly. Genome editing with engineered nucleases will likely contribute to many fields of life sciences from studying gene functions in plants and animals to gene therapy in humans. For instance, the field of synthetic biology which aims to engineer cells and organisms to perform novel functions, is likely to benefit from the ability of engineered nuclease to add or remove genomic elements and therefore create complex systems. In addition, gene functions can be studied using stem cells with engineered nucleases.

Listed below are some specific tasks this method can carry out:

- Targeted gene mutation.

- Gene therapy.
- Creating chromosome rearrangement.
- Study gene function with stem cells.
- Transgenic animals.
- Endogenous gene labeling.
- Targeted transgene addition.

The combination of recent discoveries in genetic engineering, particularly gene editing and the latest improvement in bovine reproduction technologies (e.g. *in vitro* embryo culture) allows for genome editing directly in fertilised oocytes using synthetic highly specific endonucleases. RNA-guided endonucleases:clustered regularly interspaced short palindromic repeats associated Cas9 (CRISPR/Cas9) are a new tool, further increasing the range of methods available. In particular CRISPR/Cas9 engineered endonucleases allows the use of multiple guide RNAs for simultaneous Knockouts (KO) in one step by cytoplasmic direct injection (CDI) on mammalian zygotes.

Furthermore, gene editing can be applied to certain types of fish in aquaculture such as Atlantic salmon. Gene editing in fish is currently experimental, but the possibilities include growth, disease resistance, sterility, controlled reproduction, and colour. Selecting for these traits can allow for a more sustainable environment and better welfare for the fish.

AquAdvantage salmon is a genetically modified Atlantic salmon developed by AquaBounty Technologies. The growth hormone-regulating gene in the Atlantic salmon is replaced with the growth hormone-regulating gene from the Pacific Chinook salmon and a promoter sequence from the ocean pout

Thanks to the parallel development of single-cell transcriptomics, genome editing and new stem cell models we are now entering a scientifically exciting period where functional genetics is no longer restricted to animal models but can be performed directly in human samples. Single-cell gene expression analysis has resolved a transcriptional road-map of human development from which key candidate genes are being identified for functional studies. Using global transcriptomics data to guide experimentation, the CRISPR based genome editing tool has made it feasible to disrupt or remove key genes in order to elucidate function in a human setting.

Genome editing using Meganuclease, ZFNs, and TALEN provides a new strategy for genetic manipulation in plants and are likely to assist in the engineering of desired plant traits by modifying endogenous genes. For instance, site-specific gene addition in major crop species can be used for 'trait stacking' whereby several desired traits are physically linked to ensure their co-segregation during the breeding processes. Progress in such cases have been recently reported in *Arabidopsis thaliana* and *Zea mays.* In *Arabidopsis thaliana*, using ZFN-assisted gene targeting, two herbicide-resistant genes (tobacco acetolactate synthase SuRA and SuRB) were introduced to SuR loci with as high as 2% transformed cells with mutations. In Zea mays, disruption of the target locus was achieved by ZFN-induced DSBs and the resulting NHEJ.

ZFN was also used to drive herbicide-tolerance gene expression cassette (PAT) into the targeted endogenous locus IPK1 in this case. Such genome modification observed in the regenerated plants has been shown to be inheritable and was transmitted to the next generation.

A potentially successful example of the application of genome editing techniques in crop improvement can be found in banana, where scientists used CRISPR/Cas9 editing to inactivate the endogenous banana streak virus in the B genome of banana (*Musa* spp.) to overcome a major challenge in banana breeding.

In addition, TALEN-based genome engineering has been extensively tested and optimized for use in plants. TALEN fusions have also been used by a U.S. food ingredient company, Calyxt, to improve the quality of soybean oil products and to increase the storage potential of potatoes

Several optimizations need to be made in order to improve editing plant genomes using ZFN-mediated targeting. There is a need for reliable design and subsequent test of the nucleases, the absence of toxicity of the nucleases, the appropriate choice of the plant tissue for targeting, the routes of induction of enzyme activity, the lack of off-target mutagenesis, and a reliable detection of mutated cases.

A common delivery method for CRISPR/Cas9 in plants is *Agrobacterium*-based transformation. T-DNA is introduced directly into the plant genome by a T4SS mechanism. Cas9 and gRNA-based expression cassettes are turned into Ti plasmids, which are transformed in *Agrobacterium* for plant application. To improve Cas9 delivery in live plants, viruses are being used more effective transgene delivery.

Research: Gene Therapy

The ideal gene therapy practice is that which replaces the defective gene with a normal allele at its natural location. This is advantageous over a virally delivered gene as there is no need to include the full coding sequences and regulatory sequences when only a small proportions of the gene needs to be altered as is often the case. The expression of the partially replaced genes is also more consistent with normal cell biology than full genes that are carried by viral vectors.

The first clinical use of TALEN-based genome editing was in the treatment of CD19+ acute lymphoblastic leukemia in an 11-month old child in 2015. Modified donor T cells were engineered to attack the leukemia cells, to be resistant to Alemtuzumab, and to evade detection by the host immune system after introduction.

Extensive research has been done in cells and animals using CRISPR-Cas9 to attempt to correct genetic mutations which cause genetic diseases such as Down syndrome, spina bifida, anencephaly, and Turner and Klinefelter syndromes.

In February 2019, medical scientists working with Sangamo Therapeutics, headquartered in Richmond, California, announced the first ever "in body" human gene editing therapy to permanently alter DNA - in a patient with Hunter syndrome. Clinical trials by Sangamo involving gene editing using Zinc Finger Nuclease (ZFN) are ongoing.

Research: Eradicating Diseases

Researchers have used CRISPR-Cas9 gene drives to modify genes associated with sterility in *A. gambiae*, the vector for malaria. This technique has further implications in eradicating other vector borne diseases such as yellow fever, dengue, and Zika.

The CRISPR-Cas9 system can be programmed to modulate the population of any bacterial species by targeting clinical genotypes or epidemiological isolates. It can selectively enable the beneficial bacterial species over the harmful ones by eliminating pathogen, which gives it an advantage over broad-spectrum antibiotics.

Antiviral applications for therapies targeting human viruses such as HIV, herpes, and hepatitis B virus are under research. CRISPR can be used to target the virus or the host to disrupt genes encoding the virus cell-surface receptor proteins.

In November 2018, He Jiankui announced that he had edited two human embryos, to attempt to disable the gene for CCR5, which codes for a receptor that HIV uses to enter cells. He said that twin girls, Lulu and Nana, had been born a few weeks earlier. He said that the girls still carried functional copies of CCR5 along with disabled CCR5 (mosaicism) and were still vulnerable to HIV. The work was widely condemned as unethical, dangerous, and premature.

In January 2019, scientists in China reported the creation of five identical cloned gene-edited monkeys, using the same cloning technique that was used with Zhong Zhong and Hua Hua – the first ever cloned monkeys - and Dolly the sheep, and the same gene-editing Crispr-Cas9 technique allegedly used by He Jiankui in creating the first ever gene-modified human babies Lulu and Nana. The monkey clones were made in order to study several medical diseases.

Prospects and Limitations

In the future, an important goal of research into genome editing with engineered nucleases must be the improvement of the safety and specificity of the nucleases action. For example, improving the ability to detect off-target events can improve our ability to learn about ways of preventing them. In addition, zinc-fingers used in ZFNs are seldom completely specific, and some may cause a toxic reaction. However, the toxicity has been reported to be reduced by modifications done on the cleavage domain of the ZFN.

In addition, research by Dana Carroll into modifying the genome with engineered nucleases has shown the need for better understanding of the basic recombination and repair machinery of DNA. In the future, a possible method to identify secondary targets would be to capture broken ends from cells expressing the ZFNs and to sequence the flanking DNA using high-throughput sequencing.

Because of the ease of use and cost-efficiency of CRISPR, extensive research is currently being done on it. There are now more publications on CRISPR than ZFN and TALEN despite how recent the discovery of CRISPR is. Both CRISPR and TALEN are favored to be the choices to be implemented in large-scale productions due to their precision and efficiency.

Genome editing occurs also as a natural process without artificial genetic engineering. The agents that are competent to edit genetic codes are viruses or subviral RNA-agents.

Although GEEN has higher efficiency than many other methods in reverse genetics, it is still not highly efficient; in many cases less than half of the treated populations obtain the desired changes. For example, when one is planning to use the cell's NHEJ to create a mutation, the cell's HDR systems will also be at work correcting the DSB with lower mutational rates.

Traditionally, mice have been the most common choice for researchers as a host of a disease model. CRISPR can help bridge the gap between this model and human clinical trials by creating transgenic disease models in larger animals such as pigs, dogs, and non-human primates. Using the CRISPR-Cas9 system, the programmed Cas9 protein and the sgRNA can be directly introduced into fertilized zygotes to achieve the desired gene modifications when creating transgenic models in rodents. This allows bypassing of the usual cell targeting stage in generating transgenic lines, and as a result, it reduces generation time by 90%.

One potential that CRISPR brings with its effectiveness is the application of xenotransplantation. In previous research trials, CRISPR demonstrated the ability to target and eliminate endogenous retroviruses, which reduces the risk of transmitting diseases and reduces immune barriers. Eliminating these problems improves donor organ function, which brings this application closer to a reality.

In plants, genome editing is seen as a viable solution to the conservation of biodiversity. Gene drive are a potential tool to alter the reproductive rate of invasive species, although there are significant associated risks.

Many transhumanists see genome editing as a potential tool for human enhancement. Australian biologist and Professor of Genetics David Andrew Sinclair notes that "the new technologies with genome editing will allow it to be used on individuals (...) to have (...) healthier children" – designer babies. According to a September 2016 report by the Nuffield Council on Bioethics in the future it may be possible to enhance people with genes from other organisms or wholly synthetic genes to for example improve night vision and sense of smell. George Church has compiled a list of potential genetic modifications for possibly advantageous traits such as less need for sleep, cognition-related changes that protect against Alzheimer's disease, disease resistances and enhanced learning abilities along with some of the associated studies and potential negative effects.

The American National Academy of Sciences and National Academy of Medicine issued a report in February 2017 giving qualified support to human genome editing. They recommended that clinical trials for genome editing might one day be permitted once answers have been found to safety and efficiency problems "but only for serious conditions under stringent oversight."

Risks

In the 2016 *Worldwide Threat Assessment of the US Intelligence Community* statement United States Director of National Intelligence, James R. Clapper, named genome editing as a potential weapon of mass destruction, stating that genome editing conducted by countries with regulatory or ethical standards "different from Western countries" probably increases the risk of the creation of harmful biological agents or products. According to the statement the broad distribution, low cost, and accelerated pace of development of this technology, its deliberate or unintentional misuse might lead to far-reaching economic and national security implications. For instance technologies such as CRISPR could be used to make "killer mosquitoes" that cause plagues that wipe out staple crops.

According to a September 2016 report by the Nuffield Council on Bioethics, the simplicity and low cost of tools to edit the genetic code will allow amateurs – or "biohackers" – to perform their own experiments, posing a potential risk from the release of genetically modified bugs. The review also found that the risks and benefits of modifying a person's genome – and having those changes pass on to future generations – are so complex that they demand urgent ethical scrutiny. Such modifications might have unintended consequences which could harm not only the child, but also their future children, as the altered gene would be in their sperm or eggs. In 2001 Australian researchers Ronald Jackson and Ian Ramshaw were criticized for publishing a paper in the Journal of Virology that explored the potential control of mice, a major pest in Australia, by infecting them with an altered mousepox virus that would cause infertility as the provided sensitive information could lead to the manufacture of biological weapons by potential bioterrorists who might use the knowledge to create vaccine resistant strains of other pox viruses, such as smallpox, that could affect humans. Furthermore, there are additional concerns about the ecological risks of releasing gene drives into wild populations.

Biohacking and DIY Biology

Biohacking refers to the practice of using science and technology to augment one's body and mind outside of traditional medical or scientific institutions. This can include a range of activities, from taking nootropics and experimenting with diets to more extreme practices like implanting electronic devices under the skin. The biohacking community often operates at the fringes of science, raising questions about safety, ethics, and regulation.

1.2.4 Ethical Considerations

Accessibility and Inequality

One of the primary ethical concerns with human augmentation is the potential for increased social inequality. Advanced augmentation technologies are likely to be expensive, at least initially, making them accessible primarily to the wealthy. This could exacerbate existing inequalities and create new forms of disparity between augmented and non-augmented individuals.

Identity and Humanity

Human augmentation challenges traditional notions of identity and what it means to be human. As individuals integrate more technology into their bodies, questions arise about the definition of human identity. Are individuals with significant augmentations still fully human, or do they represent a new category of being? These questions are not merely philosophical; they have practical implications for rights, responsibilities, and societal integration.

Safety and Regulation

The rapid pace of development in human augmentation technologies outstrips the current regulatory frameworks designed to ensure safety and efficacy. There is a risk of unintended consequences, both in terms of individual health and broader societal impacts. Establishing robust regulatory mechanisms that can keep pace with technological advancements is essential to mitigate these risks.

1.2.5 The Future of Human Augmentation

Integration of AI and Augmentation

The integration of artificial intelligence (AI) with augmentation technologies holds tremendous potential. AI can enhance the functionality of prosthetics, exoskeletons, and cognitive augmentation devices by providing adaptive learning and predictive capabilities. For example, AI-driven BCIs could offer real-time, personalized adjustments to cognitive enhancement techniques, optimizing performance based on individual needs and contexts.

Human Augmentation in Space Exploration

Human augmentation could play a crucial role in future space exploration. Enhancements in physical strength, endurance, and sensory perception could help astronauts withstand the rigors of space travel and adapt to extraterrestrial environments. Augmented humans might be better equipped to handle the physical and psychological challenges of long-duration missions, facilitating human presence on other planets.

Ethical and Societal Frameworks

As human augmentation technologies continue to advance, there is a pressing need for comprehensive ethical and societal frameworks. These frameworks should address issues of access, equity, identity, and safety, ensuring that the benefits of augmentation are widely shared and that potential harms are minimized. Engaging diverse stakeholders, including ethicists, scientists, policymakers, and the public, will be crucial in shaping these frameworks.

1.3 THE INTERPLAY BETWEEN HUMAN AND MACHINES

The interplay between humans and machines has been a driving force behind technological advancement and societal transformation. This dynamic relationship has evolved from simple mechanical tools to complex artificial intelligence (AI) systems, shaping the way we live, work, and interact with the world.

The interplay between humans and machines has shaped the course of technological and societal development. From early mechanical tools to advanced AI systems, this dynamic relationship has continually evolved, offering numerous benefits while posing significant challenges. As we move forward, it is essential to harness the potential of human-machine interaction in ways that enhance productivity, decision-making, and quality of life, while addressing the ethical and societal implications. By fostering collaborative intelligence and developing robust ethical frameworks, we can navigate the complexities of this relationship and ensure that the benefits of technological advancement are shared equitably across society.

1.3.1 Historical Context

Early Mechanical Tools

Forerunners of machine tools included bow drills and potter's wheels, which had existed in ancient Egypt prior to 2500 BC, and lathes, known to have existed in multiple regions of Europe since at least 1000 to 500 BC. But it was not until the later Middle Ages and the Age of Enlightenment that the modern concept of a machine tool—a class of machines used as tools in the making of metal parts, and incorporating machine-guided toolpath—began to evolve. Clockmakers of the Middle Ages and renaissance men such as Leonardo da Vinci helped expand humans' technological milieu toward the preconditions for industrial machine tools.

During the 18th and 19th centuries, and even in many cases in the 20th, the builders of machine tools tended to be the same people who would then use them to produce the end products (manufactured goods). However, from these roots also evolved an industry of machine tool builders as we define them today, meaning people who specialize in building machine tools for sale to others.

Historians of machine tools often focus on a handful of major industries that most spurred machine tool development. In order of historical emergence, they have been firearms (small arms and artillery); clocks; textile machinery; steam engines (stationary, marine, rail, and otherwise) (the story of how Watt's need for an accurate cylinder spurred Boulton's boring machine is discussed by Roe); sewing machines; bicycles; automobiles; and aircraft. Others could be included in this list as well, but they tend to be connected with the root causes already listed. For example, rolling-element bearings are an industry of themselves, but this industry's main drivers of development were the vehicles already listed—trains, bicycles, automobiles, and aircraft; and other industries, such as tractors, farm implements, and tanks, borrowed heavily from those same parent industries.

Machine tools filled a need created by textile machinery during the Industrial Revolution in England in the middle to late 1700s. Until that time, machinery was made mostly from wood, often including gearing and shafts. The increase in mechanization required more metal parts, which were usually made of cast iron or wrought iron. Cast iron could be cast in molds for larger parts, such as engine cylinders and gears, but was difficult to work with a file and could not be hammered. Red hot wrought iron could be hammered into shapes. Room temperature wrought iron was worked with a file and chisel and could be made into gears and other complex parts; however, hand working lacked precision and was a slow and expensive process.

James Watt was unable to have an accurately bored cylinder for his first steam engine, trying for several years until John Wilkinson invented a suitable boring machine in 1774, boring Boulton & Watt's first commercial engine in 1776.

The advance in the accuracy of machine tools can be traced to Henry Maudslay and refined by Joseph Whitworth. That Maudslay had established the manufacture and use of master plane gages in his shop (Maudslay & Field) located on Westminster

Road south of the Thames River in London about 1809, was attested to by James Nasmyth who was employed by Maudslay in 1829 and Nasmyth documented their use in his autobiography.

The process by which the master plane gages were produced dates back to antiquity but was refined to an unprecedented degree in the Maudslay shop. The process begins with three square plates each given an identification (ex., 1,2 and 3). The first step is to rub plates 1 and 2 together with a marking medium (called bluing today) revealing the high spots which would be removed by hand scraping with a steel scraper, until no irregularities were visible. This would not produce true plane surfaces but a "ball and socket" concave-concave and convex-convex fit, as this mechanical fit, like two perfect planes, can slide over each other and reveal no high spots. The rubbing and marking are repeated after rotating 2 relative to 1 by 90 degrees to eliminate concave-convex "potato-chip" curvature. Next, plate number 3 is compared and scraped to conform to plate number 1 in the same two trials. In this manner plates number 2 and 3 would be identical. Next plates number 2 and 3 would be checked against each other to determine what condition existed, either both plates were "balls" or "sockets" or "chips" or a combination. These would then be scraped until no high spots existed and then compared to plate number 1. Repeating this process of comparing and scraping the three plates could produce plane surfaces accurate to within millionths of an inch (the thickness of the marking medium).

The traditional method of producing the surface gages used an abrasive powder rubbed between the plates to remove the high spots, but it was Whitworth who contributed the refinement of replacing the grinding with hand scraping. Sometime after 1825, Whitworth went to work for Maudslay and it was there that Whitworth perfected the hand scraping of master surface plane gages. In his paper presented to the British Association for the Advancement of Science at Glasgow in 1840, Whitworth pointed out the inherent inaccuracy of grinding due to no control and thus unequal distribution of the abrasive material between the plates which would produce uneven removal of material from the plates.

With the creation of master plane gages of such high accuracy, all critical components of machine tools (i.e., guiding surfaces such as machine ways) could then be compared against them and scraped to the desired accuracy. The first machine tools offered for sale (i.e., commercially available) were constructed by Matthew Murray in England around 1800. Others, such as Henry Maudslay, James Nasmyth, and Joseph Whitworth, soon followed the path of expanding their entrepreneurship from manufactured end products and millwright work into the realm of building machine tools for sale.

Important early machine tools included the slide rest lathe, screw-cutting lathe, turret lathe, milling machine, pattern tracing lathe, shaper, and metal planer, which were all in use before 1840. With these machine tools the decades-old objective of producing interchangeable parts was finally realized. An important early example of something now taken for granted was the standardization of screw fasteners such as nuts and bolts. Before about the beginning of the 19th century, these were used in pairs, and even screws of the same machine were generally not interchangeable. Methods were developed to cut screw thread to a greater precision than that of the feed screw in the lathe being used. This led to the bar length standards of the 19th and early 20th centuries.

American production of machine tools was a critical factor in the Allies' victory in World War II. Production of machine tools tripled in the United States in the war. No war was more industrialized than World War II, and it has been written that the war was won as much by machine shops as by machine guns.

The production of machine tools is concentrated in about 10 countries worldwide: China, Japan, Germany, Italy, South Korea, Taiwan, Switzerland, US, Austria, Spain and a few others. Machine tool innovation continues in several public and private research centers worldwide.

Industrial Revolution

The Industrial Revolution, sometimes divided into the First Industrial Revolution and Second Industrial Revolution, was a period of global transition of the human economy towards more widespread, efficient and stable manufacturing processes that succeeded the Agricultural Revolution. Beginning in Great Britain, the Industrial Revolution spread to continental Europe and the United States, during the period from around 1760 to about 1820–1840. This transition included going from hand production methods to machines; new chemical manufacturing and iron production processes; the increasing use of water power and steam power; the development of machine tools; and the rise of the mechanized factory system. Output greatly increased, and the result was an unprecedented rise in population and the rate of population growth. The textile industry was the first to use modern production methods, and textiles became the dominant industry in terms of employment, value of output, and capital invested.

Many of the technological and architectural innovations were of British origin. By the mid-18th century, Britain was the world's leading commercial nation, controlling a global trading empire with colonies in North America and the Caribbean. Britain had major military and political hegemony on the Indian subcontinent; particularly with the proto-industrialised Mughal Bengal, through the activities of the East India Company.

The development of trade and the rise of business were among the major causes of the Industrial Revolution. Developments in law also facilitated the revolution, such as courts ruling in favour of property rights. An entrepreneurial spirit and consumer revolution helped drive industrialisation in Britain, which after 1800, was emulated in Belgium, the United States, and France.

The Industrial Revolution marked a major turning point in history, comparable only to humanity's adoption of agriculture with respect to material advancement. The Industrial Revolution influenced in some way almost every aspect of daily life. In particular, average income and population began to exhibit unprecedented sustained growth. Some economists have said the most important effect of the Industrial Revolution was that the standard of living for the general population in the Western world began to increase consistently for the first time in history, although others have said that it did not begin to improve meaningfully until the late 19th and 20th centuries. GDP per capita was broadly stable before the Industrial Revolution and the emergence of the modern capitalist economy, while the Industrial Revolution began an era of per-capita economic growth in capitalist economies. Economic historians agree that the onset of the Industrial Revolution is the most important event in human history since the domestication of animals and plants.

The precise start and end of the Industrial Revolution is still debated among historians, as is the pace of economic and social changes. According to Cambridge historian Leigh Shaw-Taylor, Britain was already industrialising in the 17th century, and "Our database shows that a groundswell of enterprise and productivity transformed the economy in the 17th century, laying the foundations for the world's first industrial economy. Britain was already a nation of makers by the year 1700" and "the history of Britain needs to be rewritten". Eric Hobsbawm held that the Industrial Revolution began in Britain in the 1780s and was not fully felt until the 1830s or 1840s, while T. S. Ashton held that it occurred roughly between 1760 and 1830. Rapid adoption of mechanized textiles spinning occurred in Britain in the 1780s, and high rates of growth in steam power and iron production occurred after 1800. Mechanized textile production spread from Great Britain to continental Europe and the United States in the early 19th century, with important centres of textiles, iron and coal emerging in Belgium and the United States and later textiles in France.

An economic recession occurred from the late 1830s to the early 1840s when the adoption of the Industrial Revolution's early innovations, such as mechanized spinning and weaving, slowed as their markets matured; and despite the increasing adoption of locomotives, steamboats and steamships, and hot blast iron smelting. New technologies such as the electrical telegraph, widely introduced in the 1840s and 1850s in the United Kingdom and the United States, were not powerful enough to drive high rates of economic growth.

Rapid economic growth began to reoccur after 1870, springing from a new group of innovations in what has been called the Second Industrial Revolution. These included new steel-making processes, mass production, assembly lines, electrical grid systems, the large-scale manufacture of machine tools, and the use of increasingly advanced machinery in steam-powered factories.

Important Technological Developments

The commencement of the Industrial Revolution is closely linked to a small number of innovations, beginning in the second half of the 18th century. By the 1830s, the following gains had been made in important technologies:

- **Textiles:** Mechanised cotton spinning powered by water, and later steam, increased the output of a worker by a factor of around 500. The power loom increased the output of a worker by a factor of over 40. The cotton gin increased productivity of removing seed from cotton by a factor of 50. Large gains in productivity also occurred in spinning and weaving of wool and linen, but they were not as great as in cotton.
- **Steam Power:** The efficiency of steam engines increased so that they used between one-fifth and one-tenth as much fuel. The adaptation of stationary steam engines to rotary motion made them suitable for industrial uses. The high-pressure engine had a high power-to-weight ratio, making it suitable for transportation. Steam power underwent a rapid expansion after 1800.
- **Iron Making:** The substitution of coke for charcoal greatly lowered the fuel cost of pig iron and wrought iron production. Using coke also allowed larger blast furnaces, resulting in economies of scale. The steam engine began being used to power blast air (indirectly by pumping water to a water wheel) in the 1750s, enabling a large increase in iron production by overcoming the limitation of water power. The cast iron blowing cylinder was first used in 1760. It was later improved by making it double acting, which allowed higher blast furnace temperatures. The puddling process produced a structural grade iron at a lower cost than the finery forge. The rolling mill was fifteen times faster than hammering wrought iron. Developed in 1828, hot blast greatly increased fuel efficiency in iron production in the following decades.
- **Invention of Machine Tools:** The first machine tools invented were the screw-cutting lathe, the cylinder boring machine, and the milling machine. Machine tools made the economical manufacture of precision metal parts possible, although it took several decades to develop

effective techniques for making interchangeable parts.

In 1750, Britain imported 2.5 million pounds of raw cotton, most of which was spun and woven by the cottage industry in Lancashire. The work was done by hand in workers' homes or occasionally in master weavers' shops. Wages in Lancashire were about six times those in India in 1770 when overall productivity in Britain was about three times higher than in India. In 1787, raw cotton consumption was 22 million pounds, most of which was cleaned, carded, and spun on machines. The British textile industry used 52 million pounds of cotton in 1800, which increased to 588 million pounds in 1850.

The share of value added by the cotton textile industry in Britain was 2.6% in 1760, 17% in 1801, and 22.4% in 1831. Value added by the British woollen industry was 14.1% in 1801. Cotton factories in Britain numbered approximately 900 in 1797. In 1760, approximately one-third of cotton cloth manufactured in Britain was exported, rising to two-thirds by 1800. In 1781, cotton spun amounted to 5.1 million pounds, which increased to 56 million pounds by 1800. In 1800, less than 0.1% of world cotton cloth was produced on machinery invented in Britain. In 1788, there were 50,000 spindles in Britain, rising to 7 million over the next 30 years.

The earliest European attempts at mechanized spinning were with wool; however, wool spinning proved more difficult to mechanize than cotton. Productivity improvement in wool spinning during the Industrial Revolution was significant but far less than that of cotton.

Arguably the first highly mechanised factory was John Lombe's water-powered silk mill at Derby, operational by 1721. Lombe learned silk thread manufacturing by taking a job in Italy and acting as an industrial spy; however, because the Italian silk industry guarded its secrets closely, the state of the industry at that time is unknown. Although Lombe's factory was technically successful, the supply of raw silk from Italy was cut off to eliminate competition. In order to promote manufacturing, the Crown paid for models of Lombe's machinery which were exhibited in the Tower of London.

Parts of India, China, Central America, South America, and the Middle East have a long history of hand manufacturing cotton textiles, which became a major industry sometime after 1000 AD. In tropical and subtropical regions where it was grown, most was grown by small farmers alongside their food crops and was spun and woven in households, largely for domestic consumption. In the 15th century, China began to require households to pay part of their taxes in cotton cloth. By the 17th century, almost all Chinese wore cotton clothing. Almost everywhere cotton cloth could be used as a medium of exchange. In India, a significant amount of cotton textiles were manufactured for distant markets, often produced by professional weavers. Some merchants also owned small weaving workshops. India produced a variety of cotton cloth, some of exceptionally fine quality.

Cotton was a difficult raw material for Europe to obtain before it was grown on colonial plantations in the Americas. The early Spanish explorers found Native Americans growing unknown species of excellent quality cotton: sea island cotton (*Gossypium barbadense*) and upland green seeded cotton *Gossypium hirsutum*. Sea island cotton grew in tropical areas and on barrier islands of Georgia and South Carolina but did poorly inland. Sea island cotton began being exported from Barbados in the 1650s. Upland green seeded cotton grew well on inland areas of the southern U.S. but was not economical because of the difficulty of removing seed, a problem solved by the cotton gin. A strain of cotton seed brought from Mexico to Natchez, Mississippi, in 1806 became the parent genetic material for over 90% of world cotton production today; it produced bolls that were three to four times faster to pick.

The Age of Discovery was followed by a period of colonialism beginning around the 16th century. Following the discovery of a trade route to India around southern Africa by the Portuguese, the British founded the East India Company, along with smaller companies of different nationalities which established trading posts and employed agents to engage in trade throughout the Indian Ocean region.

One of the largest segments of this trade was in cotton textiles, which were purchased in India and sold in Southeast Asia, including the Indonesian archipelago where spices were purchased for sale to Southeast Asia and Europe. By the mid-1760s, cloth was over three-quarters of the East India Company's exports. Indian textiles were in demand in the North Atlantic region of Europe where previously only wool and linen were available; however, the number of cotton goods consumed in Western Europe was minor until the early 19th century.

By 1600, Flemish refugees began weaving cotton cloth in English towns where cottage spinning and weaving of wool and linen was well established. They were left alone by the guilds who did not consider cotton a threat. Earlier European attempts at cotton spinning and weaving were in 12th-century Italy and 15th-century southern Germany, but these industries eventually ended when the supply of cotton was cut off. The Moors in Spain grew, spun, and wove cotton beginning around the 10th century.

British cloth could not compete with Indian cloth because India's labour cost was approximately one-fifth to one-sixth that of Britain's. In 1700 and 1721, the British government passed Calico Acts to protect the domestic woollen and linen industries from the increasing amounts of cotton fabric imported from India.

The demand for heavier fabric was met by a domestic industry based around Lancashire that produced fustian, a cloth with flax warp and cotton weft. Flax was used for the warp because wheel-spun cotton did not have sufficient strength, but the resulting blend was not as soft as 100% cotton and was more difficult to sew.

On the eve of the Industrial Revolution, spinning and weaving were done in households, for domestic consumption, and as a cottage industry under the putting-out system. Occasionally, the work was done in the workshop of a master weaver. Under the putting-out system, home-based workers produced under contract to merchant sellers, who often supplied the raw materials. In the off-season, the women, typically farmers' wives, did the spinning and the men did the weaving. Using the spinning wheel, it took anywhere from four to eight spinners to supply one handloom weaver.

The flying shuttle, patented in 1733 by John Kay—with a number of subsequent improvements including an important one in 1747—doubled the output of a weaver, worsening the imbalance between spinning and weaving. It became widely used around Lancashire after 1760 when John's son, Robert, invented the dropbox, which facilitated changing thread colors.

Lewis Paul patented the roller spinning frame and the flyer-and-bobbin system for drawing wool to a more even thickness. The technology was developed with the help of John Wyatt of Birmingham. Paul and Wyatt opened a mill in Birmingham which used their rolling machine powered by a donkey. In 1743, a factory opened in Northampton with 50 spindles on each of five of Paul and Wyatt's machines. This operated until about 1764. A similar mill was built by Daniel Bourn in Leominster, but this burnt down. Both Lewis Paul and Daniel Bourn patented carding machines in 1748. Based on two sets of rollers that travelled at different speeds, it was later used in the first cotton spinning mill.

In 1764, in the village of Stanhill, Lancashire, James Hargreaves invented the spinning jenny, which he patented in 1770. It was the first practical spinning frame with multiple spindles. The jenny worked in a similar manner to the spinning wheel, by first clamping down on the fibres, then by drawing them out, followed by twisting. It was a simple, wooden framed machine that only cost about £6 for a 40-spindle model in 1792 and was used mainly by home spinners. The jenny produced a lightly twisted yarn only suitable for weft, not warp.

The spinning frame or water frame was developed by Richard Arkwright who, along with two partners, patented it in 1769. The design was partly based on a spinning machine built by Kay, who was hired by Arkwright. For each spindle the water frame used a series of four pairs of rollers, each operating at a successively higher rotating speed, to draw out the fibre which was then twisted by the spindle. The roller spacing was slightly longer than the fibre length. Too close a spacing caused the fibres to break while too distant a spacing caused uneven thread. The top rollers were leather-covered and loading on the rollers was applied by a weight.

The weights kept the twist from backing up before the rollers. The bottom rollers were wood and metal, with fluting along the length. The water frame was able to produce a hard, medium-count thread suitable for warp, finally allowing 100% cotton cloth to be made in Britain. Arkwright and his partners used water power at a factory in Cromford, Derbyshire in 1771, giving the invention its name.

Samuel Crompton invented the spinning mule in 1779, so called because it is a hybrid of Arkwright's water frame and James Hargreaves's spinning jenny in the same way that a mule is the product of crossbreeding a female horse with a male donkey. Crompton's mule was able to produce finer thread than hand spinning and at a lower cost. Mule-spun thread was of suitable strength to be used as a warp and finally allowed Britain to produce highly competitive yarn in large quantities.

Realising that the expiration of the Arkwright patent would greatly increase the supply of spun cotton and lead to a shortage of weavers, Edmund Cartwright developed a vertical power loom which he patented in 1785. In 1776, he patented a two-man operated loom. Cartwright's loom design had several flaws, the most serious being thread breakage. Samuel Horrocks patented a fairly successful loom in 1813. Horock's loom was improved by Richard Roberts in 1822, and these were produced in large numbers by Roberts, Hill & Co. Roberts was additionally a maker of high-quality machine tools and a pioneer in the use of jigs and gauges for precision workshop measurement.

The demand for cotton presented an opportunity to planters in the Southern United States, who thought upland cotton would be a profitable crop if a better way could be found to remove the seed. Eli Whitney responded to the challenge by inventing the inexpensive cotton gin. A man using a cotton gin could remove seed from as much upland cotton in one day as would previously have taken two months to process, working at the rate of one pound of cotton per day.

These advances were capitalised on by entrepreneurs, of whom the best known is Arkwright. He is credited with a list of inventions, but these were actually developed by such people as Kay and Thomas Highs; Arkwright nurtured the inventors, patented the ideas, financed the initiatives, and protected the machines. He created the cotton mill which brought the production processes together in a factory, and he developed the use of power—first horsepower and then water power—which made cotton manufacture a mechanised industry. Other inventors increased the efficiency of the individual steps of spinning (carding, twisting and spinning, and rolling) so that the supply of yarn increased greatly. Steam power was then applied to drive textile machinery. Manchester acquired the nickname Cottonopolis during the early 19th century owing to its sprawl of textile factories.

Although mechanization dramatically decreased the cost of cotton cloth, by the mid-19th century machine-woven cloth still could not equal the quality of hand-woven Indian cloth, in part because of the fineness of thread made possible by the type of cotton used in India, which allowed high thread counts. However, the high productivity of British textile manufacturing allowed coarser grades of British cloth to undersell hand-spun and woven fabric in low-wage India, eventually destroying the Indian industry.

Bar iron was the commodity form of iron used as the raw material for making hardware goods such as nails, wire, hinges, horseshoes, wagon tires, chains, etc., as well as structural shapes. A small amount of bar iron was converted into steel. Cast iron was used for pots, stoves, and other items where its brittleness was tolerable. Most cast iron was refined and converted to bar iron, with substantial losses. Bar iron was made by the bloomery process, which was the predominant iron smelting process until the late 18th century.

In the UK in 1720, there were 20,500 tons of cast iron produced with charcoal and 400 tons with coke. In 1750 charcoal iron production was 24,500 and coke iron was 2,500 tons. In 1788, the production of charcoal cast iron was 14,000 tons while coke iron production was 54,000 tons. In 1806, charcoal cast iron production was 7,800 tons and coke cast iron was 250,000 tons.

In 1750, the UK imported 31,200 tons of bar iron and either refined from cast iron or directly produced 18,800 tons of bar iron using charcoal and 100 tons using coke. In 1796, the UK was making 125,000 tons of bar iron with coke and 6,400 tons with charcoal; imports were 38,000 tons and exports were 24,600 tons. In 1806 the UK did not import bar iron but exported 31,500 tons.

A major change in the iron industries during the Industrial Revolution was the replacement of wood and other bio-fuels with coal; for a given amount of heat, mining coal required much less labour than cutting wood and converting it to charcoal, and coal was much more abundant than wood, supplies of which were becoming scarce before the enormous increase in iron production that took place in the late 18th century.

In 1709, Abraham Darby made progress using coke to fuel his blast furnaces at Coalbrookdale. However, the coke pig iron he made was not suitable for making wrought iron and was used mostly for the production of cast iron goods, such as pots and kettles. He had the advantage over his rivals in that his pots, cast by his patented process, were thinner and cheaper than theirs.

In 1750, coke had generally replaced charcoal in the smelting of copper and lead and was in widespread use in glass production. In the smelting and refining of iron, coal and coke produced inferior iron to that made with charcoal because of the coal's sulfur content. Low sulfur coals were known, but they still contained harmful amounts.

Conversion of coal to coke only slightly reduces the sulfur content. A minority of coals are coking. Another factor limiting the iron industry before the Industrial Revolution was the scarcity of water power to power blast bellows. This limitation was overcome by the steam engine.

Use of coal in iron smelting started somewhat before the Industrial Revolution, based on innovations by Clement Clerke and others from 1678, using coal reverberatory furnaces known as cupolas. These were operated by the flames playing on the ore and charcoal or coke mixture, reducing the oxide to metal. This has the advantage that impurities (such as sulphur ash) in the coal do not migrate into the metal. This technology was applied to lead from 1678 and to copper from 1687. It was also applied to iron foundry work in the 1690s, but in this case the reverberatory furnace was known as an air furnace. (The foundry cupola is a different, and later, innovation.)

Coke pig iron was hardly used to produce wrought iron until 1755–56, when Darby's son Abraham Darby II built furnaces at Horsehay and Ketley where low sulfur coal was available (and not far from Coalbrookdale). These furnaces were equipped with water-powered bellows, the water being pumped by Newcomen steam engines. The Newcomen engines were not attached directly to the blowing cylinders because the engines alone could not produce a steady air blast. Abraham Darby III installed similar steam-pumped, water-powered blowing cylinders at the Dale Company when he took control in 1768. The Dale Company used several Newcomen engines to drain its mines and made parts for engines which it sold throughout the country.

Steam engines made the use of higher pressure and volume blast practical; however, the leather used in bellows was expensive to replace. In 1757, ironmaster John Wilkinson patented a hydraulic powered blowing engine for blast furnaces. The blowing cylinder for blast furnaces was introduced in 1760 and the first blowing cylinder made of cast iron is believed to be the one used at Carrington in 1768 that was designed by John Smeaton.

Cast iron cylinders for use with a piston were difficult to manufacture; the cylinders had to be free of holes and had to be machined smooth and straight to remove any warping. James Watt had great difficulty trying to have a cylinder made for his first steam engine. In 1774 Wilkinson invented a precision boring machine for boring cylinders. After Wilkinson bored the first successful cylinder for a Boulton and Watt steam engine in 1776, he was given an exclusive contract for providing cylinders. After Watt developed a rotary steam engine in 1782, they were widely applied to blowing, hammering, rolling and slitting.

The solutions to the sulfur problem were the addition of sufficient limestone to the furnace to force sulfur into the slag as well as the use of low sulfur coal. The use of lime or limestone required higher furnace temperatures to form a free-flowing slag. The increased furnace temperature made possible by improved blowing also increased the capacity of blast furnaces and allowed for increased furnace height. In addition to lower cost and greater availability, coke had other important advantages over charcoal in that it was harder and made the column of materials (iron ore, fuel, slag) flowing down the blast furnace more porous and did not crush in the much taller furnaces of the late 19th century.

As cast iron became cheaper and widely available, it began being a structural material for bridges and buildings. A famous early example is the Iron Bridge built in 1778 with cast iron produced by Abraham Darby III. However, most cast iron was converted to wrought iron. Conversion of cast iron had long been done in a finery forge. An improved refining process known as potting and stamping was developed, but this was superseded by Henry Cort's puddling process. Cort developed two significant iron manufacturing processes: rolling in 1783 and puddling in 1784. Puddling produced a structural grade iron at a relatively low cost.

Puddling was a means of decarburizing molten pig iron by slow oxidation in a reverberatory furnace by manually stirring it with a long rod. The decarburized iron, having a higher melting point than cast iron, was raked into globs by the puddler. When the glob was large enough, the puddler would remove it. Puddling was backbreaking and extremely hot work. Few puddlers lived to be 40. Because puddling was done in a reverberatory furnace, coal or coke could be used as fuel. The puddling process continued to be used until the late 19th century when iron was being displaced by steel.

Because puddling required human skill in sensing the iron globs, it was never successfully mechanised. Rolling was an important part of the puddling process because the grooved rollers expelled most of the molten slag and consolidated the mass of hot wrought iron. Rolling was 15 times faster at this than a trip hammer. A different use of rolling, which was done at lower temperatures than that for expelling slag, was in the production of iron sheets, and later structural shapes such as beams, angles, and rails.

The puddling process was improved in 1818 by Baldwyn Rogers, who replaced some of the sand lining on the reverberatory furnace bottom with iron oxide. In 1838 John Hall patented the use of roasted tap cinder (iron silicate) for the furnace bottom, greatly reducing the loss of iron through increased slag caused by a sand lined bottom. The tap cinder also tied up some phosphorus, but this was not understood at the time. Hall's process also used iron scale or rust which reacted with carbon in the molten iron. Hall's process, called *wet puddling*, reduced losses of iron with the slag from almost 50% to around 8%.

Puddling became widely used after 1800. Up to that time, British iron manufacturers had used considerable amounts of iron imported from Sweden and Russia to supplement domestic supplies. Because of the increased British production, imports began to decline in 1785, and by the 1790s Britain eliminated imports and became a net exporter of bar iron.

Hot blast, patented by the Scottish inventor James Beaumont Neilson in 1828, was the most important development of the 19th century for saving energy in making pig iron. By using preheated combustion air, the amount of fuel to make a unit of pig iron was reduced at first by between one-third using coke or two-thirds using coal; the efficiency gains continued as the technology improved. Hot blast also raised the operating temperature of furnaces, increasing their capacity. Using less coal or coke meant introducing fewer impurities into the pig iron. This meant that lower quality coal could be used in areas where coking coal was unavailable or too expensive; however, by the end of the 19th century transportation costs fell considerably.

Shortly before the Industrial Revolution, an improvement was made in the production of steel, which was an expensive commodity and used only where iron would not do, such as for cutting edge tools and for springs. Benjamin Huntsman developed his crucible steel technique in the 1740s. The raw material for this was blister steel, made by the cementation process. The supply of cheaper iron and steel aided a number of industries, such as those making nails, hinges, wire, and other hardware items. The development of machine tools allowed better working of iron, causing it to be increasingly used in the rapidly growing machinery and engine industries.

The development of the stationary steam engine was an important element of the Industrial Revolution; however, during the early period of the Industrial Revolution, most industrial power was supplied by water and wind. In Britain, by 1800 an estimated 10,000 horsepower was being supplied by steam. By 1815 steam power had grown to 210,000 hp.

The first commercially successful industrial use of steam power was patented by Thomas Savery in 1698. He constructed in London a low-lift combined vacuum and pressure water pump that generated about one horsepower (hp) and was used in numerous waterworks and in a few mines (hence its "brand name", *The Miner's Friend*). Savery's pump was economical in small horsepower ranges but was prone to boiler explosions in larger sizes. Savery pumps continued to be produced until the late 18th century.

The first successful piston steam engine was introduced by Thomas Newcomen before 1712. Newcomen engines were installed for draining hitherto unworkable deep mines, with the engine on the surface; these were large machines, requiring a significant amount of capital to build, and produced upwards of 3.5 kW (5 hp).

They were also used to power municipal water supply pumps. They were extremely inefficient by modern standards, but when located where coal was cheap at pit heads, they opened up a great expansion in coal mining by allowing mines to go deeper. Despite their disadvantages, Newcomen engines were reliable and easy to maintain and continued to be used in the coalfields until the early decades of the 19th century.

By 1729, when Newcomen died, his engines had spread to Hungary in 1722, and then to Germany, Austria, and Sweden. A total of 110 are known to have been built by 1733 when the joint patent expired, of which 14 were abroad. In the 1770s the engineer John Smeaton built some very large examples and introduced a number of improvements. A total of 1,454 engines had been built by 1800.

A fundamental change in working principles was brought about by Scotsman James Watt. With financial support from his business partner Englishman Matthew Boulton, he had succeeded by 1778 in perfecting his steam engine, which incorporated a series of radical improvements, notably the closing off of the upper part of the cylinder thereby making the low-pressure steam drive the top of the piston instead of the atmosphere; use of a steam jacket; and the celebrated separate steam condenser chamber. The separate condenser did away with the cooling water that had been injected directly into the cylinder which cooled the cylinder and wasted steam. Likewise, the steam jacket kept steam from condensing in the cylinder, also improving efficiency. These improvements increased engine efficiency so that Boulton and Watt's engines used only 20–25% as much coal per horsepower-hour as Newcomen's. Boulton and Watt opened the Soho Foundry for the manufacture of such engines in 1795.

In 1783, the Watt steam engine had been fully developed into a double-acting rotative type, which meant that it could be used to directly drive the rotary machinery of a factory or mill. Both of Watt's basic engine types were commercially very successful, and by 1800 the firm Boulton & Watt had constructed 496 engines, with 164 driving reciprocating pumps, 24 serving blast furnaces, and 308 powering mill machinery; most of the engines generated from 3.5 to 7.5 kW (5 to 10 hp).

Until about 1800, the most common pattern of steam engine was the beam engine, built as an integral part of a stone or brick engine-house, but soon various patterns of self-contained rotative engines (readily removable but not on wheels) were developed, such as the table engine. Around the start of the 19th century, at which time the Boulton and Watt patent expired, the Cornish engineer Richard Trevithick and the American Oliver Evans began to construct higher-pressure non-condensing steam engines, exhausting against the atmosphere. High pressure yielded an engine and boiler compact enough to be used on mobile road and rail locomotives and steamboats.

Small industrial power requirements continued to be provided by animal and human muscle until widespread electrification in the early 20th century. These included crank-powered, treadle-powered and horse-powered workshop, and light industrial machinery.

Pre-industrial machinery was built by various craftsmen—millwrights built watermills and windmills; carpenters made wooden framing; and smiths and turners made metal parts. Wooden components had the disadvantage of changing dimensions with temperature and humidity, and the various joints tended to rack (work loose) over time. As the Industrial Revolution progressed, machines with metal parts and frames became more common. Other important uses of metal parts were in firearms and threaded fasteners, such as machine screws, bolts, and nuts. There was also the need for precision in making parts. Precision would allow better working machinery, interchangeability of parts, and standardization of threaded fasteners.

The demand for metal parts led to the development of several machine tools. They have their origins in the tools developed in the 18th century by makers of clocks and watches and scientific instrument makers to enable them to batch-produce small mechanisms. Before the advent of machine tools, metal was worked manually using the basic hand tools of hammers, files, scrapers, saws, and chisels. Consequently, the use of metal machine parts was kept to a minimum. Hand methods of production were laborious and costly, and precision was difficult to achieve.

The first large precision machine tool was the cylinder boring machine invented by John Wilkinson in 1774. It was designed to bore the large cylinders on early steam engines. Wilkinson's machine was the first to use the principle of line-boring, where the tool is supported on both ends, unlike earlier designs used for boring cannon that relied on a less stable cantilevered boring bar.

The planing machine, the milling machine and the shaping machine were developed in the early decades of the 19th century. Although the milling machine was invented at this time, it was not developed as a serious workshop tool until somewhat later in the 19th century.James Fox of Derby and Matthew Murray of Leeds were manufacturers of machine tools who found success in exporting from England and are also notable for having developed the planer around the same time as Richard Roberts of Manchester.

Henry Maudslay, who trained a school of machine tool makers early in the 19th century, was a mechanic with superior ability who had been employed at the Royal Arsenal, Woolwich. He worked as an apprentice at the Royal Arsenal under Jan Verbruggen. In 1774 Verbruggen had installed a horizontal boring machine which was the first industrial size lathe in the UK. Maudslay was hired away by Joseph

Bramah for the production of high-security metal locks that required precision craftsmanship. Bramah patented a lathe that had similarities to the slide rest lathe. Maudslay perfected the slide rest lathe, which could cut machine screws of different thread pitches by using changeable gears between the spindle and the lead screw. Before its invention, screws could not be cut to any precision using various earlier lathe designs, some of which copied from a template. The slide rest lathe was called one of history's most important inventions. Although it was not entirely Maudslay's idea, he was the first person to build a functional lathe using a combination of known innovations of the lead screw, slide rest, and change gears.

Maudslay left Bramah's employment and set up his own shop. He was engaged to build the machinery for making ships' pulley blocks for the Royal Navy in the Portsmouth Block Mills. These machines were all-metal and were the first machines for mass production and making components with a degree of interchangeability. The lessons Maudslay learned about the need for stability and precision he adapted to the development of machine tools, and in his workshops, he trained a generation of men to build on his work, such as Richard Roberts, Joseph Clement and Joseph Whitworth.

The techniques to make mass-produced metal parts of sufficient precision to be interchangeable is largely attributed to a program of the U.S. Department of War which perfected interchangeable parts for firearms in the early 19th century. In the half-century following the invention of the fundamental machine tools, the machine industry became the largest industrial sector of the U.S. economy, by value added.

The large-scale production of chemicals was an important development during the Industrial Revolution. The first of these was the production of sulphuric acid by the lead chamber process invented by the Englishman John Roebuck (James Watt's first partner) in 1746. He was able to greatly increase the scale of the manufacture by replacing the relatively expensive glass vessels formerly used with larger, less expensive chambers made of riveted sheets of lead. Instead of making a small amount each time, he was able to make around 50 kilograms (100 pounds) in each of the chambers, at least a tenfold increase.

The production of an alkali on a large scale became an important goal as well, and Nicolas Leblanc succeeded in 1791 in introducing a method for the production of sodium carbonate (soda ash). The Leblanc process was a reaction of sulfuric acid with sodium chloride to give sodium sulfate and hydrochloric acid. The sodium sulfate was heated with calcium carbonate and coal to give a mixture of sodium carbonate and calcium sulfide. Adding water separated the soluble sodium carbonate from the calcium sulfide. The process produced a large amount of pollution (the hydrochloric acid was initially vented to the atmosphere, and calcium sulfide was a waste product). Nonetheless, this synthetic soda ash proved economical compared to that produced from burning specific plants (barilla or kelp), which were the previously dominant sources of soda ash, and also to potash (potassium carbonate) produced from hardwood ashes. These two chemicals were very important because they enabled the introduction of a host of other inventions, replacing many small-scale operations with more cost-effective and controllable processes. Sodium carbonate had many uses in the glass, textile, soap, and paper industries. Early uses for sulfuric acid included pickling (removing rust from) iron and steel, and for bleaching cloth.

The development of bleaching powder (calcium hypochlorite) by Scottish chemist Charles Tennant in about 1800, based on the discoveries of French chemist Claude Louis Berthollet, revolutionised the bleaching processes in the textile industry by dramatically reducing the time required (from months to days) for the traditional process then in use, which required repeated exposure to the sun in bleach fields after soaking the textiles with alkali or sour milk. Tennant's factory at St Rollox, Glasgow, became the largest chemical plant in the world.

After 1860 the focus on chemical innovation was in dyestuffs, and Germany took world leadership, building a strong chemical industry. Aspiring chemists flocked to German universities in the 1860–1914 era to learn the latest techniques. British scientists by contrast, lacked research universities and did not train advanced students; instead, the practice was to hire German-trained chemists.

In 1824 Joseph Aspdin, a British bricklayer turned builder, patented a chemical process for making portland cement which was an important advance in the building trades.

This process involves sintering a mixture of clay and limestone to about 1,400 °C (2,552 °F), then grinding it into a fine powder which is then mixed with water, sand and gravel to produce concrete. Portland cement concrete was used by the English engineer Marc Isambard Brunel several years later when constructing the Thames Tunnel. Concrete was used on a large scale in the construction of the London sewer system a generation later.

Though others made a similar innovation elsewhere, the large-scale introduction of gas lighting was the work of William Murdoch, an employee of Boulton & Watt. The process consisted of the large-scale gasification of coal in furnaces, the purification of the gas (removal of sulphur, ammonia, and heavy hydrocarbons), and its storage and distribution.

The first gas lighting utilities were established in London between 1812 and 1820. They soon became one of the major consumers of coal in the UK. Gas lighting affected social and industrial organisation because it allowed factories and stores to remain open longer than with tallow candles or oil lamps. Its introduction allowed nightlife to flourish in cities and towns as interiors and streets could be lighted on a larger scale than before.

Glass was made in ancient Greece and Rome. A new method of glass production, known as the cylinder process, was developed in Europe during the early 19th century. In 1832 this process was used by the Chance Brothers to create sheet glass. They became the leading producers of window and plate glass.

This advancement allowed for larger panes of glass to be created without interruption, thus freeing up the space planning in interiors as well as the fenestration of buildings. The Crystal Palace is the supreme example of the use of sheet glass in a new and innovative structure.

A machine for making a continuous sheet of paper on a loop of wire fabric was patented in 1798 by Louis-Nicolas Robert in France. The paper machine is known as a Fourdrinier after the financiers, brothers Sealy and Henry Fourdrinier, who were stationers in London.

Although greatly improved and with many variations, the Fourdrinier machine is the predominant means of paper production today. The method of continuous production demonstrated by the paper machine influenced the development of continuous rolling of iron and later steel and other continuous production processes.

The British Agricultural Revolution is considered one of the causes of the Industrial Revolution because improved agricultural productivity freed up workers to work in other sectors of the economy. In contrast, per-capita food supply in Europe was stagnant or declining and did not improve in some parts of Europe until the late 18th century.

The English lawyer Jethro Tull invented an improved seed drill in 1701. It was a mechanical seeder that distributed seeds evenly across a plot of land and planted them at the correct depth. This was important because the yield of seeds harvested to seeds planted at that time was around four or five. Tull's seed drill was very expensive and not very reliable and therefore did not have much of an effect. Good quality seed drills were not produced until the mid 18th century.

Joseph Foljambe's *Rotherham plough* of 1730 was the first commercially successful iron plough. The threshing machine, invented by the Scottish engineer Andrew Meikle in 1784, displaced hand threshing with a flail, a laborious job that took about one-quarter of agricultural labour. Lower labor requirements subsequently result in lowered wages and numbers of farm labourers, who faced near starvation, leading to the 1830 agricultural rebellion of the Swing Riots.

Machine tools and metalworking techniques developed during the Industrial Revolution eventually resulted in precision manufacturing techniques in the late

19th century for mass-producing agricultural equipment, such as reapers, binders, and combine harvesters.

Coal mining in Britain, particularly in South Wales, started early. Before the steam engine, pits were often shallow bell pits following a seam of coal along the surface, which were abandoned as the coal was extracted. In other cases, if the geology was favourable the coal was mined by means of an adit or drift mine driven into the side of a hill. Shaft mining was done in some areas, but the limiting factor was the problem of removing water. It could be done by hauling buckets of water up the shaft or to a sough (a tunnel driven into a hill to drain a mine). In either case, the water had to be discharged into a stream or ditch at a level where it could flow away by gravity.

The introduction of the steam pump by Thomas Savery in 1698 and the Newcomen steam engine in 1712 greatly facilitated the removal of water and enabled shafts to be made deeper, enabling more coal to be extracted.

These were developments that had begun before the Industrial Revolution, but the adoption of John Smeaton's improvements to the Newcomen engine followed by James Watt's more efficient steam engines from the 1770s reduced the fuel costs of engines, making mines more profitable. The Cornish engine, developed in the 1810s, was much more efficient than the Watt steam engine.

Coal mining was very dangerous owing to the presence of firedamp in many coal seams. Some degree of safety was provided by the safety lamp which was invented in 1816 by Sir Humphry Davy and independently by George Stephenson. However, the lamps proved a false dawn because they became unsafe very quickly and provided a weak light.

Firedamp explosions continued, often setting off coal dust explosions, so casualties grew during the entire 19th century. Conditions of work were very poor, with a high casualty rate from rock falls.

At the beginning of the Industrial Revolution, inland transport was by navigable rivers and roads, with coastal vessels employed to move heavy goods by sea. Wagonways were used for conveying coal to rivers for further shipment, but canals had not yet been widely constructed.

Animals supplied all of the motive power on land, with sails providing the motive power on the sea. The first horse railways were introduced toward the end of the 18th century, with steam locomotives being introduced in the early decades of the 19th century. Improving sailing technologies boosted average sailing speed by 50% between 1750 and 1830.

The Industrial Revolution improved Britain's transport infrastructure with a turnpike road network, a canal and waterway network, and a railway network. Raw materials and finished products could be moved more quickly and cheaply than before. Improved transportation also allowed new ideas to spread quickly.

Before and during the Industrial Revolution navigation on several British rivers was improved by removing obstructions, straightening curves, widening and deepening, and building navigation locks. Britain had over 1,600 kilometres (1,000 mi) of navigable rivers and streams by 1750. Canals and waterways allowed bulk materials to be economically transported long distances inland. This was because a horse could pull a barge with a load dozens of times larger than the load that could be drawn in a cart.

Canals began to be built in the UK in the late 18th century to link the major manufacturing centres across the country. Known for its huge commercial success, the Bridgewater Canal in North West England, which opened in 1761 and was mostly funded by The 3rd Duke of Bridgewater. From Worsley to the rapidly growing town of Manchester its construction cost £168,000 (£22,589,130 as of 2013), but its advantages over land and river transport meant that within a year of its opening in 1761, the price of coal in Manchester fell by about half. This success helped inspire a period of intense canal building, known as Canal Mania. Canals were hastily built with the aim of replicating the commercial success of the Bridgewater Canal, the most notable being the Leeds and Liverpool Canal and the Thames and Severn Canal which opened in 1774 and 1789 respectively.

By the 1820s a national network was in existence. Canal construction served as a model for the organisation and methods later used to construct the railways. They were eventually largely superseded as profitable commercial enterprises by the spread of the railways from the 1840s on. The last major canal to be built in the United Kingdom was the Manchester Ship Canal, which upon opening in 1894 was the largest ship canal in the world, and opened Manchester as a port. However, it never achieved the commercial success its sponsors had hoped for and signalled canals as a dying mode of transport in an age dominated by railways, which were quicker and often cheaper.

Britain's canal network, together with its surviving mill buildings, is one of the most enduring features of the early Industrial Revolution to be seen in Britain.

France was known for having an excellent system of roads at the time of the Industrial Revolution; however, most of the roads on the European continent and in the UK were in bad condition and dangerously rutted.

Much of the original British road system was poorly maintained by thousands of local parishes, but from the 1720s (and occasionally earlier) turnpike trusts were set up to charge tolls and maintain some roads. Increasing numbers of main roads were turnpiked from the 1750s to the extent that almost every main road in England and Wales was the responsibility of a turnpike trust.

New engineered roads were built by John Metcalf, Thomas Telford and most notably John McAdam, with the first 'macadam' stretch of road being Marsh Road at Ashton Gate, Bristol in 1816. The first macadam road in the U.S. was the "Boonsborough Turnpike Road" between Hagerstown and Boonsboro, Maryland in 1823.

The major turnpikes radiated from London and were the means by which the Royal Mail was able to reach the rest of the country. Heavy goods transport on these roads was by means of slow, broad-wheeled carts hauled by teams of horses. Lighter goods were conveyed by smaller carts or by teams of packhorse.

Stagecoaches carried the rich, and the less wealthy could pay to ride on carriers carts. Productivity of road transport increased greatly during the Industrial Revolution, and the cost of travel fell dramatically. Between 1690 and 1840 productivity almost tripled for long-distance carrying and increased four-fold in stage coaching.

Railways were made practical by the widespread introduction of inexpensive puddled iron after 1800, the rolling mill for making rails, and the development of the high-pressure steam engine also around 1800. Reducing friction was one of the major reasons for the success of railroads compared to wagons. This was demonstrated on an iron plate-covered wooden tramway in 1805 at Croydon, England.

A good horse on an ordinary turnpike road can draw two thousand pounds, or one ton. A party of gentlemen were invited to witness the experiment, that the superiority of the new road might be established by ocular demonstration. Twelve wagons were loaded with stones, till each wagon weighed three tons, and the wagons were fastened together.

A horse was then attached, which drew the wagons with ease, six miles [10 km] in two hours, having stopped four times, in order to show he had the power of starting, as well as drawing his great load.

Wagonways for moving coal in the mining areas had started in the 17th century and were often associated with canal or river systems for the further movement of coal. These were all horse-drawn or relied on gravity, with a stationary steam engine to haul the wagons back to the top of the incline. The first applications of the steam locomotive were on wagon or plate ways (as they were then often called from the cast-iron plates used). Horse-drawn public railways begin in the early 19th century when improvements to pig and wrought iron production were lowering costs.

The Digital Age

The mid-20th century saw the rise of computers, ushering in the digital age. The development of electronic computers revolutionized data processing and information management. Human-machine interaction became more sophisticated, with the introduction of graphical user interfaces (GUIs) and programming languages that allowed users to communicate more effectively with machines.

1.3.2 Modern Human-Machine Interaction

Artificial Intelligence and Machine Learning

Artificial intelligence and machine learning represent the latest frontier in human-machine interaction.

These technologies enable machines to learn from data, make decisions, and perform tasks that previously required human intelligence. AI-powered systems are now used in various fields, including healthcare, finance, transportation, and entertainment.

Robotics and Automation

Robotics and automation have transformed industries by taking over repetitive and hazardous tasks.

Industrial robots are used in manufacturing to assemble products with precision and speed. In addition, service robots are being developed for applications such as healthcare, hospitality, and domestic chores, enhancing efficiency and safety.

Human-Computer Interfaces

Advancements in human-computer interfaces (HCIs) have made interactions with machines more intuitive and seamless. Touchscreens, voice recognition, and gesture control are now commonplace, allowing users to interact with devices in natural ways.

Virtual reality (VR) and augmented reality (AR) are further pushing the boundaries of HCI by creating immersive and interactive environments.

1.3.3 Benefits of Human-Machine Interaction

Increased Productivity and Efficiency

One of the most significant benefits of human-machine interaction is increased productivity and efficiency. Machines can perform tasks faster and with greater accuracy than humans, leading to higher output and reduced errors.

Automation in manufacturing, for example, has drastically improved production rates and product quality.

Enhanced Decision-Making

AI and machine learning enhance decision-making by providing insights from vast amounts of data. Predictive analytics, for instance, can help businesses forecast market trends and consumer behavior, enabling more informed decisions.

In healthcare, AI can analyze medical records and imaging data to assist doctors in diagnosing diseases and planning treatments.

Improved Quality of Life

Machines and AI technologies have the potential to improve the quality of life for individuals. Assistive technologies, such as prosthetics and exoskeletons, enhance mobility for people with disabilities. Smart home devices can automate household tasks, providing convenience and freeing up time for other activities.

1.3.4 Challenges and Ethical Considerations

Job Displacement

One of the primary challenges of human-machine interaction is job displacement. As machines and AI take over tasks traditionally performed by humans, there is a risk of widespread unemployment, particularly in industries reliant on manual labor. This raises concerns about economic inequality and the need for retraining and upskilling programs to help workers transition to new roles.

Privacy and Security

The integration of AI and machines into various aspects of life raises significant privacy and security concerns. AI systems often require access to large amounts of personal data, which can be vulnerable to breaches and misuse. Ensuring robust data protection measures and ethical guidelines for data usage is crucial to safeguarding individual privacy.

Ethical Decision-Making

As machines become more autonomous, ethical decision-making becomes a critical issue. AI systems used in healthcare, law enforcement, and finance, for instance, must make decisions that align with ethical standards and societal values. Ensuring that these systems are transparent, fair, and accountable is essential to prevent biases and unintended consequences.

1.3.5 The Future of Human-Machine Interaction

Human Augmentation

The future of human-machine interaction includes the possibility of human augmentation, where technology enhances human physical and cognitive abilities. Wearable devices, brain-computer interfaces, and genetic engineering are areas of research that could lead to significant enhancements in human capabilities, potentially blurring the line between human and machine.

Collaborative Intelligence

The concept of collaborative intelligence emphasizes the complementary strengths of humans and machines. Rather than replacing humans, machines can augment human abilities, leading to more effective collaboration. In this model, humans provide creativity, empathy, and strategic thinking, while machines offer computational power, data analysis, and repetitive task automation.

Ethical Frameworks and Governance

As human-machine interaction continues to evolve, developing robust ethical frameworks and governance structures will be critical. Policymakers, technologists, and ethicists must work together to create guidelines that ensure the responsible development and deployment of AI and machine technologies. These frameworks should address issues such as transparency, accountability, fairness, and the long-term impact on society.

CHAPTER 2

FOUNDATIONS OF HUMAN AUGMENTATION

2.1 COGNITIVE AUGMENTATION: ENHANCING MENTAL CAPABILITIES

Cognitive augmentation refers to the use of technology and techniques to enhance human mental capabilities. This burgeoning field combines advances in neuroscience, artificial intelligence, and bioengineering to improve memory, attention, creativity, and other cognitive functions. As society increasingly relies on information and complex decision-making, cognitive augmentation holds the promise of significantly impacting education, work, and everyday life.

Cognitive augmentation represents a frontier in human enhancement, offering the potential to significantly improve mental capabilities. From nootropics and brain stimulation to brain-computer interfaces and neurofeedback, a range of technologies and techniques are being explored to enhance cognitive function. While the benefits of cognitive augmentation are substantial, including improved learning, workplace performance, and mental health, there are also significant ethical and societal considerations to address.

Ensuring equity, safety, informed consent, and data privacy are crucial as these technologies develop. As we look to the future, the integration of AI and advances in neurotechnology will likely drive further innovations, making cognitive augmentation an integral part of enhancing human potential.

By developing robust ethical frameworks and fostering inclusive access, we can navigate the complexities of cognitive augmentation and harness its benefits for the betterment of society.

2.1.1 Methods and Technologies for Cognitive Augmentation

Nootropics and Cognitive Enhancers

Nootropics, also known as smart drugs, are substances designed to improve cognitive function. These can range from natural supplements like caffeine and ginkgo biloba to prescription medications such as modafinil and methylphenidate. Nootropics are believed to enhance cognitive abilities by increasing neurotransmitter levels, improving blood flow to the brain, and protecting neurons from damage.

Brain Stimulation Techniques

Brain stimulation techniques, such as transcranial magnetic stimulation (TMS) and transcranial direct current stimulation (tDCS), have gained attention for their potential to enhance cognitive functions. TMS uses magnetic fields to stimulate nerve cells in the brain, while tDCS delivers a low electrical current to specific brain areas. These non-invasive methods have shown promise in improving memory, learning, and problem-solving skills.

Brain-Computer Interfaces (BCIs)

Brain-computer interfaces (BCIs) enable direct communication between the brain and external devices. BCIs can interpret brain signals and translate them into commands for computers or prosthetic devices. This technology has the potential to enhance cognitive capabilities by providing new ways to interact with information and control devices, particularly for individuals with disabilities.

Neurofeedback

Neurofeedback is a form of biofeedback that uses electrical potentials in the brain to reinforce desired brain states through operant conditioning. This process is non-invasive and typically collects brain activity data using electroencephalography (EEG). Several neurofeedback protocols exist, with potential additional benefit from use of quantitative electroencephalography (QEEG) or functional magnetic resonance imaging (fMRI) to localize and personalize treatment. Related technologies include functional near-infrared spectroscopy-mediated (fNIRS) neurofeedback, hemoencephalography biofeedback (HEG), and fMRI biofeedback.

Placebo-controlled trials have often found the control group to show the same level of improvement as the group receiving actual neurofeedback treatment, which suggests these improvements may be caused by secondary effects instead. Neurofeedback has been shown to trigger positive behavioral outcomes, such as relieving symptoms related to psychiatric disorders or improving specific cognitive functions in healthy participants. These positive behavioral outcomes rely on brain plasticity mechanisms and the ability of subjects to learn throughout life.

History

In 1898, Edward Thorndike formulated the law of effect. In his work, he theorized that behavior is shaped by satisfying or discomforting consequences. This set the foundation for operant conditioning. In 1924, the German psychiatrist Hans Berger connected several electrodes to a patient's scalp and detected a small current by using a ballistic galvanometer. In his subsequent studies, Berger analyzed EEGs qualitatively, but in 1932, G. Dietsch applied Fourier analysis to seven EEG records and later became the first researcher to apply quantitative EEG (QEEG).

In 1950, Neal E. Miller of Yale University was able to train mice to regulate their heartbeat frequency. Later on, he continued his work with humans, training them through auditory feedback.

The first study to demonstrate neurofeedback was reported by Joe Kamiya in 1962. Kamiya's experiment had two parts: In the first part, a subject was asked to keep their eyes closed, and when a tone sounded, to say whether they were experiencing alpha waves. Initially, the subject would guess correctly about fifty percent of the time, but some subjects would eventually develop the ability to better distinguish between states.

M. Barry Sterman trained cats to modify their EEG patterns to exhibit more of the so-called sensorimotor rhythm (SMR). He published this research in 1967. Sterman subsequently discovered that the SMR-trained cats were much more resistant to epileptic seizures after exposure to the convulsant chemical monomethylhydrazine than non-trained cats. In 1971, he reported similar improvements with an epileptic patient whose seizures could be controlled through SMR training. Joel Lubar contributed to the research of EEG biofeedback, starting with epilepsy and later with hyperactivity and ADHD. Ming-Yang Cheng was instrumental in advancing research on EEG neurofeedback, specifically targeting enhancements in SMR power among skilled golfers.

Neuroplasticity

In 2010, a study provided some evidence of neuroplastic changes occurring after brainwave training. In this study, half an hour of voluntary control of brain rhythms led to a lasting shift in cortical excitability and intracortical function. The authors observed that the cortical response to transcranial magnetic stimulation (TMS) was significantly enhanced after neurofeedback, persisted for at least twenty minutes, and was correlated with an EEG time-course indicative of activity-dependent plasticity

Types of Neurofeedback

The term neurofeedback is not legally protected. There are various approaches that give feedback about neuronal activity, and as such are referred to as "neurofeedback" by their respective operators. Distinctions can be made on several levels. The first takes into account which technology is being used (EEG, fMRI, fNIRS, HEG). Nonetheless, further distinctions are crucial even within the realm of EEG neurofeedback, as different methodologies of analysis can be chosen, some of which are backed up by a higher number of peer-reviewed studies, whereas for others, scientific literature is scarce, and explanatory models are entirely missing.

Despite these differences, a common denominator can be found in the requirement of providing feedback. Usually, feedback is provided by auditory or visual input.

While original feedback was provided by sounding tones according to neurological activity, many new ways have been found. It is possible to listen to music or podcasts where the volume is controlled as feedback, for example. Often, visual feedback is used in the form of animations on a TV screen. Visual feedback can also be provided in combination with videos and films, or even during reading tasks where the brightness of the screen represents the direct feedback. Simple games can also be used, where the game itself is controlled by the brain activity. Recent developments have tried to incorporate virtual reality (VR), and controllers can already be used for more involved engagement with the feedback.

Amplitude training, or frequency band training (used synonymously), is the method with the largest body of scientific literature; it also represents the original method of EEG neurofeedback. The EEG signal is analyzed with respect to its frequency spectrum, split into the common frequency bands used in EEG neuroscience (delta, theta, alpha, beta, gamma). The activity involves training the amplitude of a certain frequency band on a defined location on the scalp to higher or lower values.

Depending on the training goal (for example, increasing attention and focus, reaching a calm state, reducing epileptic seizures, etc.), the electrodes have to be placed in different positions. Additionally, the trained frequency bands and the training directions (to higher or lower amplitudes) might vary according to the training goal. Thus, EEG wave components that are expected to be beneficial to the training goal are rewarded with positive feedback when appearing and/or increasing in amplitude. Frequency band amplitudes that are expected to be hindering are trained downwards by reinforcement through the feedback.

As an example, considering ADHD, this would result in training low-beta or mid-beta frequencies in the central-to-frontal lobe to increase in amplitude, while simultaneously trying to reduce theta and high-beta amplitudes in the same region of the brain.

In the sports domain, SMR training has garnered attention, with a substantial body of research suggesting that enhancing it could improve performance. This improvement is particularly evident after multiple training sessions designed to enhance motor skills critical for precise movements. Such precision is required in various sports activities, including golf putting, soccer free kicks, and basketball free throws.

For SCP (slow cortical potentials) training, one trains the DC voltage component of the EEG signal. The application of this type of EEG neurofeedback training was mostly endorsed by research done by Niels Birbaumer and his group. The most common symptom base for SCP training is ADHD, whereas SCPs also find their application in brain-computer interfaces.

Normal EEG signals are restricted to the surface of the scalp. Using a high number of electrodes (19 or more), the source of certain electrical events can be localized. Similar to a tomography that renders a 3D image out of many 2D images, the many EEG channels are used to create LORETA images that represent in 3D the electrical activity distribution within the brain.

The LORETA method can be used in combination with MRI to merge structural and functional activities. It is able to provide even better temporal resolution than PET or fMRI. For the application with live neurofeedback, however, 19-channel neurofeedback and LORETA has limited scientific evidence, and until now, shows no benefit over traditional 1- or 2-channel neurofeedback.

Discussion and Critique

There is ongoing discussion about the effect size of neurofeedback in the scientific literature. As neurofeedback is explained mostly based on the model of operant conditioning, the sensitivity of the feedback (the difficulty to receive a reward) also plays a role. It has been shown that the desired conditioning can be reversed if threshold values are set too low. Other publications have not found any effect of neurofeedback, apart from placebo, when using automatic thresholds that update every thirty seconds in order to maintain a constant success rate of 80%.

2.1.2 Potential Benefits of Cognitive Augmentation

Improved Learning and Education

Cognitive augmentation technologies can revolutionize education by enhancing learning capabilities and personalizing educational experiences. Techniques like neurofeedback and brain stimulation can improve focus and memory retention, allowing students to grasp complex concepts more easily. AI-driven educational tools can adapt to individual learning styles and provide targeted support, making education more effective and inclusive.

Enhanced Workplace Performance

In the workplace, cognitive augmentation can boost productivity and creativity. Nootropics and brain stimulation can enhance concentration and problem-solving skills, enabling employees to perform complex tasks more efficiently. BCIs and AI tools can assist in data analysis and decision-making, reducing cognitive load and allowing workers to focus on strategic thinking and innovation.

Better Mental Health and Well-being

Cognitive augmentation also holds promise for improving mental health and overall well-being. Techniques like neurofeedback can help manage stress and anxiety, while brain stimulation has shown potential in treating depression and other mental health disorders. By enhancing cognitive resilience and emotional regulation, these technologies can contribute to a better quality of life.

2.1.3 Ethical Considerations

Equity and Accessibility

One of the primary ethical concerns with cognitive augmentation is equity and accessibility. Advanced cognitive enhancement technologies may be expensive and inaccessible to many, leading to a potential cognitive divide. Ensuring that these technologies are affordable and widely available is crucial to prevent exacerbating existing social inequalities.

Safety and Long-term Effects

The safety and long-term effects of cognitive augmentation technologies are not yet fully understood. Nootropics and brain stimulation methods, for example, may have side effects or unknown long-term consequences. Rigorous scientific research and clinical trials are necessary to establish the safety and efficacy of these interventions before they are widely adopted.

Ethical Use and Consent

The ethical use of cognitive augmentation technologies requires informed consent and respect for individual autonomy. Users must be fully aware of the potential risks and benefits of these technologies and make informed decisions about their use. Additionally, there are concerns about the potential misuse of cognitive augmentation in competitive environments, such as academic testing or professional settings, where individuals might feel pressured to enhance their cognitive abilities artificially.

Privacy and Data Security

Cognitive augmentation technologies, particularly BCIs and AI-driven tools, involve the collection and analysis of vast amounts of personal data. Ensuring the privacy and security of this data is paramount to prevent misuse and protect individual rights. Robust data protection measures and ethical guidelines must be established to safeguard against potential breaches and abuses.

2.1.4 Future Directions

Integration of AI and Cognitive Augmentation

The integration of artificial intelligence with cognitive augmentation technologies holds significant potential for enhancing mental capabilities. AI can provide personalized recommendations for cognitive enhancement strategies, optimize brain stimulation protocols, and analyze neurofeedback data to improve outcomes.

This synergy between AI and cognitive augmentation can lead to more effective and tailored interventions.

Advances in Neurotechnology

Continued advances in neurotechnology will likely drive the future of cognitive augmentation. Developments in brain imaging, neural interfaces, and bioelectronics will provide deeper insights into brain function and enable more precise and effective augmentation techniques. Innovations such as neural dust and nanoscale brain-machine interfaces could revolutionize the way we interact with and enhance our cognitive capabilities.

Societal and Ethical Frameworks

As cognitive augmentation technologies evolve, there is a pressing need for comprehensive societal and ethical frameworks. These frameworks should address issues of access, equity, safety, and ethical use, ensuring that the benefits of cognitive augmentation are shared broadly and responsibly. Engaging diverse stakeholders, including ethicists, policymakers, scientists, and the public, will be crucial in shaping these frameworks.

Enhancing Human Potential

Ultimately, the goal of cognitive augmentation is to enhance human potential and improve quality of life. By augmenting mental capabilities, individuals can achieve greater heights in education, work, and personal development. However, it is essential to balance the pursuit of enhancement with ethical considerations, ensuring that cognitive augmentation contributes to a more equitable and just society.

2.2 PHYSICAL AUGMENTATION: BEYOND HUMAN STRENGTH AND ENDURANCE

Physical augmentation refers to the enhancement of human physical capabilities through technological, biological, and mechanical means. This field aims to transcend the natural limitations of human strength, endurance, and overall physical performance, providing numerous benefits in healthcare, sports, the military, and everyday life.

Physical augmentation represents a frontier in human enhancement, offering the potential to significantly improve physical capabilities. From exoskeletons and advanced prosthetics to genetic and biomedical enhancements, a range of technologies and techniques are being explored to enhance physical performance. While the benefits of physical augmentation are substantial, including improved mobility, enhanced occupational performance, and accelerated healing, there are also significant ethical and societal considerations to address. Ensuring equity, safety, informed consent, and data privacy are crucial as these technologies develop. As we look to the future, the integration of AI and advances in biotechnology will likely drive further innovations, making physical augmentation an integral part of enhancing human potential. By developing robust ethical frameworks and fostering inclusive access, we can navigate the complexities of physical augmentation and harness its benefits for the betterment of society.

2.2.1 Methods and Technologies for Physical Augmentation

Exoskeletons

Exoskeletons are wearable robotic devices that enhance the physical capabilities of the wearer. These devices can be powered or passive, providing support and increased strength to the user. Powered exoskeletons use motors and sensors to assist movement, allowing users to lift heavier loads and move with greater ease. They are used in various fields, including industrial work, military operations, and rehabilitation for individuals with mobility impairments.

Prosthetics and Bionics

Prosthetics and bionics are artificial devices that replace missing or damaged body parts, restoring function and, in some cases, providing enhanced capabilities. Modern prosthetics are often equipped with advanced sensors and actuators that allow for more natural and precise movements. Bionic limbs, which are controlled by the user's neural signals, offer an even higher level of integration and functionality, enabling users to perform complex tasks with improved dexterity and strength.

Genetic and Biomedical Enhancements

Genetic and biomedical enhancements involve altering the human body at the cellular or genetic level to improve physical capabilities. Techniques such as gene editing (e.g., CRISPR) can potentially enhance muscle growth, increase bone density, and improve endurance by modifying specific genes.

Additionally, biomedical enhancements like tissue engineering and regenerative medicine aim to repair or replace damaged tissues, leading to improved physical performance and recovery from injuries.

Performance-Enhancing Drugs

Performance-enhancing drugs (PEDs) are substances that improve physical performance by increasing strength, endurance, and recovery. These can include anabolic steroids, growth hormones, and other substances that boost muscle mass and enhance athletic performance. While PEDs are often associated with sports and bodybuilding, they also have applications in medical treatments and rehabilitation.

2.2.2 Benefits of Physical Augmentation

Improved Mobility and Independence

Physical augmentation technologies can significantly improve mobility and independence for individuals with disabilities or injuries. Exoskeletons and advanced prosthetics enable users to walk, lift objects, and perform daily activities with greater ease, enhancing their quality of life. These technologies also play a crucial role in rehabilitation, helping patients regain strength and function after injury or surgery.

Enhanced Occupational Performance

In industrial and military settings, physical augmentation can enhance occupational performance by reducing physical strain and increasing productivity. Exoskeletons, for example, can help workers lift heavy objects and perform repetitive tasks without experiencing fatigue or injury. In the military, augmented soldiers can carry heavier loads and endure longer missions, improving overall effectiveness and safety.

Advances in Sports and Athletics

Physical augmentation has the potential to revolutionize sports and athletics by pushing the boundaries of human performance. Enhanced training methods, advanced prosthetics, and performance-enhancing drugs can help athletes achieve new records and overcome physical limitations. These advancements also open up new possibilities for inclusive sports, allowing athletes with disabilities to compete at the highest levels.

Accelerated Healing and Recovery

Biomedical enhancements and regenerative medicine can accelerate healing and recovery from injuries. Techniques such as stem cell therapy, gene editing, and tissue engineering can repair damaged tissues and improve the body's natural healing processes. This can lead to faster recovery times and better outcomes for patients, reducing the long-term impact of injuries and illnesses.

2.2.3 Challenges of Physical Augmentation

Ethical and Fairness Issues

One of the primary challenges of physical augmentation is addressing ethical and fairness issues, particularly in sports and competitive environments. The use of performance-enhancing drugs and advanced prosthetics can create an uneven playing field, raising questions about fairness and the integrity of competition. Establishing clear guidelines and regulations is essential to ensure fair competition and ethical use of augmentation technologies.

Safety and Long-term Effects

The safety and long-term effects of physical augmentation technologies are not yet fully understood. Performance-enhancing drugs, for example, can have significant side effects and long-term health consequences. Similarly, the use of exoskeletons and advanced prosthetics may pose risks if not properly designed and tested. Rigorous research and clinical trials are necessary to establish the safety and efficacy of these technologies.

Accessibility and Cost

Ensuring accessibility and affordability of physical augmentation technologies is a significant challenge. Advanced prosthetics, exoskeletons, and genetic enhancements can be expensive, limiting access for individuals who could benefit from them. Developing cost-effective solutions and implementing policies to promote equitable access are crucial to preventing disparities and ensuring that these technologies benefit a wide range of people.

Social and Psychological Impact

The widespread use of physical augmentation technologies may have profound social and psychological impacts. The potential for enhanced physical capabilities could lead to societal pressure to augment, creating a divide between augmented and non-augmented individuals. Additionally, the psychological effects of using augmentation technologies, such as changes in self-perception and identity, need to be carefully considered and addressed.

2.2.4 Ethical Considerations

Informed Consent and Autonomy

Ensuring informed consent and respecting individual autonomy are fundamental ethical principles in physical augmentation. Individuals must be fully aware of the potential risks and benefits of augmentation technologies and be able to make informed decisions about their use. Transparent communication and education about these technologies are crucial for maintaining autonomy and trust.

Equity and Access

To prevent exacerbating social inequalities, it is essential to ensure that physical augmentation technologies are accessible to all, regardless of socioeconomic status. This requires addressing the affordability of these technologies and implementing policies that promote equitable access. Ensuring that physical enhancement is not limited to a privileged few is critical for fostering an inclusive society.

Safety and Efficacy

The safety and efficacy of physical augmentation technologies must be rigorously tested and validated through scientific research and clinical trials. Understanding the long-term effects and potential risks is essential to prevent harm and ensure that these technologies are safe for widespread use. Regulatory frameworks should be established to oversee the development and deployment of augmentation technologies.

Privacy and Data Protection

Given the vast amounts of personal data involved in physical augmentation, robust data protection measures are essential to safeguard privacy. Ethical guidelines and regulations must be in place to ensure that data is collected, stored, and used responsibly. Protecting individual privacy and preventing misuse of data are critical ethical considerations in the development of augmentation technologies.

2.2.5 Future Directions

Integration of AI and Physical Augmentation

The integration of artificial intelligence (AI) with physical augmentation technologies holds significant potential for enhancing physical capabilities. AI can optimize the performance of exoskeletons and prosthetics, providing real-time adjustments and feedback to improve functionality. Additionally, AI-driven diagnostics and personalized treatment plans can enhance the effectiveness of biomedical enhancements.

Advances in Biotechnology

Continued advances in biotechnology will likely drive the future of physical augmentation. Developments in gene editing, tissue engineering, and regenerative medicine will provide new opportunities for enhancing physical capabilities and treating a wide range of conditions. Innovations such as lab-grown organs and personalized medicine could revolutionize healthcare and physical augmentation.

Societal and Ethical Frameworks

As physical augmentation technologies evolve, there is a pressing need for comprehensive societal and ethical frameworks. These frameworks should address issues of access, equity, safety, and ethical use, ensuring that the benefits of physical augmentation are shared broadly and responsibly. Engaging diverse stakeholders, including ethicists, policymakers, scientists, and the public, will be crucial in shaping these frameworks.

Enhancing Human Potential

Ultimately, the goal of physical augmentation is to enhance human potential and improve quality of life. By augmenting physical capabilities, individuals can achieve greater heights in various fields, from healthcare and sports to everyday activities. However, it is essential to balance the pursuit of enhancement with ethical considerations, ensuring that physical augmentation contributes to a more equitable and just society.

2.3 SENSORY AUGMENTATION: EXPANDING PERCEPTUAL HORIZONS

Sensory augmentation refers to the enhancement or extension of human sensory capabilities through technology. This rapidly evolving field leverages advancements in neuroscience, biotechnology, and engineering to enhance or create new senses, allowing individuals to perceive and interact with the world in unprecedented ways. Sensory augmentation holds significant potential for improving quality of life, expanding human experiences, and offering new opportunities for individuals with sensory impairments.

Sensory augmentation represents a frontier in human enhancement, offering the potential to significantly improve sensory capabilities. From auditory and visual augmentation to tactile feedback and sensory substitution, a range of technologies and techniques are being explored to enhance perception. While the benefits of sensory augmentation are substantial, including enhanced quality of life, expanded human experiences, and improved communication, significant ethical and societal considerations must be addressed. Ensuring equity, safety, informed consent, and data privacy are crucial as these technologies develop. As we look to the future, the integration of AI and advances in neuroscience and biotechnology will likely drive further innovations, making sensory augmentation an integral part of enhancing human potential. By developing robust ethical frameworks and fostering inclusive access, we can navigate the complexities of sensory augmentation and harness its benefits for the betterment of society.

2.3.1 Methods and Technologies for Sensory Augmentation

Auditory Augmentation

Auditory augmentation includes technologies designed to enhance or restore hearing. Hearing aids and cochlear implants are well-known devices that improve auditory perception for individuals with hearing loss. Hearing aids amplify sound, while cochlear implants bypass damaged parts of the ear and directly stimulate the auditory nerve, enabling sound perception. Advances in auditory augmentation are also exploring ways to enhance natural hearing capabilities, such as expanding the range of frequencies that can be heard.

Visual Augmentation

Visual augmentation technologies aim to enhance or restore vision. Eyeglasses and contact lenses are common forms of visual augmentation that correct refractive errors. More advanced technologies include retinal implants and bionic eyes, which can restore vision to individuals with retinal damage by converting light into electrical signals that the brain can interpret. Augmented reality (AR) and virtual reality (VR) systems also enhance visual experiences by overlaying digital information onto the physical world or creating entirely immersive environments.

Tactile Augmentation

Tactile augmentation involves enhancing the sense of touch. Haptic feedback devices are used in various applications, from virtual reality systems to medical training simulators, to provide realistic tactile sensations. Wearable haptic devices can convey information through vibrations or pressure, improving interactions with digital environments. Additionally, advanced prosthetics equipped with sensory feedback systems allow users to experience touch sensations, improving the functionality and integration of artificial limbs.

Olfactory and Gustatory Augmentation

Olfactory and gustatory augmentation focuses on enhancing the senses of smell and taste. While this area is less developed than auditory and visual augmentation, there are ongoing efforts to create devices that can simulate or enhance olfactory and gustatory experiences. For example, electronic noses and tongues can detect and analyze chemical compounds, potentially aiding individuals with impaired senses of smell or taste. Such technologies could also be used in virtual reality systems to create more immersive experiences.

Sensory Substitution

Sensory substitution involves replacing one sensory modality with another. This technique is often used to assist individuals with sensory impairments. For example, the BrainPort device converts visual information into tactile sensations on the tongue, enabling blind individuals to "see" through touch. Similarly, the vOICe system translates visual images into soundscapes, allowing users to perceive their environment through hearing. These technologies leverage the brain's plasticity, enabling it to interpret new types of sensory input.

2.3.2 Benefits of Sensory Augmentation

Enhanced Quality of Life

Sensory augmentation technologies can significantly enhance the quality of life for individuals with sensory impairments. Devices like cochlear implants and retinal implants can restore lost senses, enabling individuals to participate more fully in social, educational, and professional activities. Enhanced sensory capabilities also improve safety and independence, reducing reliance on others for assistance.

Expanded Human Experiences

Sensory augmentation can expand human experiences by enhancing or creating new senses. For example, augmented reality devices can overlay additional information onto the physical world, enriching the user's experience and providing new ways to interact with their environment.

Sensory augmentation can also enable the perception of previously imperceptible phenomena, such as electromagnetic fields or ultrasonic sounds, broadening the scope of human perception.

Improved Communication and Interaction

Enhanced sensory capabilities can improve communication and interaction. For example, advanced hearing aids with directional microphones and noise-canceling features enable clearer conversations in noisy environments. Tactile feedback devices can provide more intuitive interactions with digital interfaces, enhancing user experience in various applications, from gaming to remote surgery.

Scientific and Technological Advancements

The development of sensory augmentation technologies drives scientific and technological advancements. Research into the brain's plasticity and sensory processing mechanisms contributes to our understanding of human perception and cognition. Additionally, innovations in materials science, bioengineering, and computational neuroscience are essential for creating effective sensory augmentation devices, leading to broader applications and new technological breakthroughs.

2.3.3 Challenges of Sensory Augmentation

Technical Limitations

Technical limitations are a significant challenge in the development of sensory augmentation technologies. Creating devices that accurately mimic or enhance natural sensory experiences is complex and requires advanced engineering and neuroscience knowledge. Ensuring the reliability, durability, and user-friendliness of these devices is crucial for their widespread adoption.

Cost and Accessibility

The cost of sensory augmentation technologies can be prohibitive, limiting access for many individuals who could benefit from them. Advanced devices like cochlear implants and retinal implants are expensive, and their affordability is a significant barrier to widespread use. Developing cost-effective solutions and implementing policies to promote equitable access are essential to ensure that these technologies benefit a broad range of people.

Adaptation and Learning

Users of sensory augmentation devices often require a period of adaptation and learning to effectively use the new sensory input. This can be a challenging and time-consuming process, particularly for individuals with significant sensory impairments. Providing adequate training and support is crucial to ensure successful integration and use of these technologies.

Informed Consent and Autonomy

Ensuring informed consent and respecting individual autonomy are fundamental ethical principles in sensory augmentation. Users must be fully aware of the potential risks and benefits of augmentation technologies and make informed decisions about their use. Transparent communication and education about these technologies are crucial for maintaining autonomy and trust.

Equity and Access

To prevent exacerbating social inequalities, it is essential to ensure that sensory augmentation technologies are accessible to all, regardless of socioeconomic status. Addressing the affordability of these technologies and implementing policies that promote equitable access are critical to fostering an inclusive society. Ensuring that sensory enhancement is not limited to a privileged few is essential for social equity.

Privacy and Data Protection

Given the personal data involved in sensory augmentation, robust data protection measures are essential to safeguard privacy. Ethical guidelines and regulations must ensure that data is collected, stored, and used responsibly. Protecting individual privacy and preventing data misuse are critical ethical considerations in developing sensory augmentation technologies.

Long-term Effects and Safety

The long-term effects and safety of sensory augmentation technologies are not yet fully understood. Ensuring the safety and efficacy of these devices through rigorous scientific research and clinical trials is crucial. Understanding the potential risks and long-term impacts is essential to prevent harm and ensure these technologies are safe for widespread use.

2.3.4 Future Directions

Integration of AI and Sensory Augmentation

The integration of artificial intelligence (AI) with sensory augmentation technologies holds significant potential for enhancing sensory capabilities. AI can optimize the performance of sensory devices, providing real-time adjustments and personalized feedback to improve functionality. AI-driven algorithms can also analyze sensory data to enhance perception and provide new ways of interacting with the environment.

Advances in Neuroscience and Biotechnology

Continued advances in neuroscience and biotechnology will drive the future of sensory augmentation. Developments in neural interfaces, bioelectronics, and materials science will provide new opportunities for creating more effective and sophisticated sensory devices. Innovations such as neural dust and nanoscale brain-machine interfaces could revolutionize sensory augmentation by providing more seamless integration with the nervous system.

Societal and Ethical Frameworks

As sensory augmentation technologies evolve, there is a pressing need for comprehensive societal and ethical frameworks. These frameworks should address issues of access, equity, safety, and ethical use, ensuring that the benefits of sensory augmentation are shared broadly and responsibly. Engaging diverse stakeholders, including ethicists, policymakers, scientists, and the public, will be crucial in shaping these frameworks.

Enhancing Human Potential

Ultimately, the goal of sensory augmentation is to enhance human potential and improve the quality of life. By augmenting sensory capabilities, individuals can achieve greater heights in various fields, from healthcare and education to everyday activities. Balancing the pursuit of enhancement with ethical considerations is essential to ensure that sensory augmentation contributes to a more equitable and just society.

CHAPTER 3
HUMAN-MACHINE INTERFACES (HMIS)

3.1 BRAIN-COMPUTER INTERFACES (BCIS)

A brain–computer interface (BCI), sometimes called a brain–machine interface (BMI), is a direct communication link between the brain's electrical activity and an external device, most commonly a computer or robotic limb. BCIs are often directed at researching, mapping, assisting, augmenting, or repairing human cognitive or sensory-motor functions.

They are often conceptualized as a human–machine interface that skips the intermediary of moving body parts (hands...), although they also raise the possibility of erasing the distinction between brain and machine. BCI implementations range from non-invasive (EEG, MEG, MRI) and partially invasive (ECoG and endovascular) to invasive (microelectrode array), based on how physically close electrodes are to brain tissue.

Research on BCIs began in the 1970s by Jacques Vidal at the University of California, Los Angeles (UCLA) under a grant from the National Science Foundation, followed by a contract from DARPA. Vidal's 1973 paper introduced the expression *brain-computer interface* into scientific literature.

Due to the cortical plasticity of the brain, signals from implanted prostheses can, after adaptation, be handled by the brain like natural sensor or effector channels. Following years of animal experimentation, the first neuroprosthetic devices were implanted in humans in the mid-1990s.

Studies in human-computer interaction via the application of machine learning to statistical temporal features extracted from the frontal lobe (EEG brainwave) data has achieved success in classifying mental states (relaxed, neutral, concentrating), mental emotional states (negative, neutral, positive), and thalamocortical dysrhythmia.

3.1.1 History

The history of brain-computer interfaces (BCIs) starts with Hans Berger's discovery of the brain's electrical activity and the development of electroencephalography (EEG). In 1924 Berger was the first to record human brain activity utilizing EEG.

Berger was able to identify oscillatory activity, such as the alpha wave (8–13 Hz), by analyzing EEG traces.

Berger's first recording device was rudimentary. He inserted silver wires under the scalps of his patients. These were later replaced by silver foils attached to the patient's head by rubber bandages. Berger connected these sensors to a Lippmann capillary electrometer, with disappointing results. However, more sophisticated measuring devices, such as the Siemens double-coil recording galvanometer, which displayed voltages as small as 10 volt, led to success.

Berger analyzed the interrelation of alternations in his EEG wave diagrams with brain diseases. EEGs permitted completely new possibilities for brain research.

Although the term had not yet been coined, one of the earliest examples of a working brain-machine interface was the piece *Music for Solo Performer* (1965) by American composer Alvin Lucier. The piece makes use of EEG and analog signal processing hardware (filters, amplifiers, and a mixing board) to stimulate acoustic percussion instruments. Performing the piece requires producing alpha waves and thereby "playing" the various instruments via loudspeakers that are placed near or directly on the instruments.

Vidal coined the term "BCI" and produced the first peer-reviewed publications on this topic. He is widely recognized as the inventor of BCIs. A review pointed out that Vidal's 1973 paper stated the "BCI challenge" of controlling external objects using EEG signals, and especially use of Contingent Negative Variation (CNV) potential as a challenge for BCI control. Vidal's 1977 experiment was the first application of BCI after his 1973 BCI challenge. It was a noninvasive EEG (actually Visual Evoked Potentials (VEP)) control of a cursor-like graphical object on a computer screen. The demonstration was movement in a maze.

1988 was the first demonstration of noninvasive EEG control of a physical object, a robot. The experiment demonstrated EEG control of multiple start-stop-restart cycles of movement, along an arbitrary trajectory defined by a line drawn on a floor. The line-following behavior was the default robot behavior, utilizing autonomous intelligence and an autonomous energy source.

In 1990, a report was given on a closed loop, bidirectional, adaptive BCI controlling a computer buzzer by an anticipatory brain potential, the Contingent Negative Variation (CNV) potential. The experiment described how an expectation state of the brain, manifested by CNV, used a feedback loop to control the S2 buzzer in the S1-S2-CNV paradigm.

The resulting cognitive wave representing the expectation learning in the brain was termed Electroexpectogram (EXG). The CNV brain potential was part of Vidal's 1973 challenge.

Studies in the 2010s suggested neural stimulation's potential to restore functional connectivity and associated behaviors through modulation of molecular mechanisms. This opened the door for the concept that BCI technologies may be able to restore function.

Beginning in 2013, DARPA funded BCI technology through the BRAIN initiative, which supported work out of teams including University of Pittsburgh Medical Center, Paradromics, Brown, and Synchron.

3.1.2 Neuroprosthetics

Neuroprosthetics is an area of neuroscience concerned with neural prostheses, that is, using artificial devices to replace the function of impaired nervous systems and brain-related problems, or of sensory or other organs (bladder, diaphragm, etc.). As of December 2010, cochlear implants had been implanted as neuroprosthetic devices in some 736,900 people worldwide. Other neuroprosthetic devices aim to restore vision, including retinal implants. The first neuroprosthetic device, however, was the pacemaker.

The terms are sometimes used interchangeably. Neuroprosthetics and BCIs seek to achieve the same aims, such as restoring sight, hearing, movement, ability to communicate, and even cognitive function. Both use similar experimental methods and surgical techniques.

3.1.3 Animal research

Several laboratories have managed to read signals from monkey and rat cerebral cortices to operate BCIs to produce movement. Monkeys have moved computer cursors and commanded robotic arms to perform simple tasks simply by thinking about the task and seeing the results, without motor output. In May 2008 photographs that showed a monkey at the University of Pittsburgh Medical Center operating a robotic arm by thinking were published in multiple studies. Sheep have also been used to evaluate BCI technology including Synchron's Stentrode.

In 2020, Elon Musk's Neuralink was successfully implanted in a pig. In 2021, Musk announced that the company had successfully enabled a monkey to play video games using Neuralink's device.

Early work

In 1969 operant conditioning studies by Fetz et.al. at the Regional Primate Research Center and Department of Physiology and Biophysics, University of Washington School of Medicine showed that monkeys could learn to control the deflection of a biofeedback arm with neural activity. Similar work in the 1970s established that monkeys could learn to control the firing rates of individual and multiple neurons in the primary motor cortex if they were rewarded accordingly.

Algorithms to reconstruct movements from motor cortex neurons, which control movement, date back to the 1970s. In the 1980s, Georgopoulos at Johns Hopkins University found a mathematical relationship between the electrical responses of single motor cortex neurons in rhesus macaque monkeys and the direction in which they moved their arms. He also found that dispersed groups of neurons, in different areas of the monkey's brains, collectively controlled motor commands. He was able to record the firings of neurons in only one area at a time, due to equipment limitations.

Several groups have been able to capture complex brain motor cortex signals by recording from neural ensembles (groups of neurons) and using these to control external devices.

Research

Kennedy and Yang Dan

Phillip Kennedy (Neural Signals founder (1987) and colleagues built the first intracortical brain–computer interface by implanting neurotrophic-cone electrodes into monkeys. In 1999, Yang Dan et.al. at University of California, Berkeley decoded neuronal firings to reproduce images from cats. The team used an array of electrodes embedded in the thalamus (which integrates the brain's sensory input). Researchers targeted 177 brain cells in the thalamus lateral geniculate nucleus area, which decodes signals from the retina. Neuron firings were recorded from watching eight short movies. Using mathematical filters, the researchers decoded the signals to reconstruct recognizable scenes and moving objects.

Nicolelis

Duke University professor Miguel Nicolelis advocates using multiple electrodes spread over a greater area of the brain to obtain neuronal signals.

After initial studies in rats during the 1990s, Nicolelis and colleagues developed BCIs that decoded brain activity in owl monkeys and used the devices to reproduce monkey movements in robotic arms. Monkeys' advanced reaching and grasping abilities and hand manipulation skills, made them good test subjects.

By 2000, the group succeeded in building a BCI that reproduced owl monkey movements while the monkey operated a joystick or reached for food. The BCI operated in real time and could remotely control a separate robot. But the monkeys received no feedback (open-loop BCI).

Later experiments on rhesus monkeys included feedback and reproduced monkey reaching and grasping movements in a robot arm. Their deeply cleft and furrowed brains made them better models for human neurophysiology than owl monkeys. The monkeys were trained to reach and grasp objects on a computer screen by manipulating a joystick while corresponding movements by a robot arm were hidden. The monkeys were later shown the robot and learned to control it by viewing its movements. The BCI used velocity predictions to control reaching movements and simultaneously predicted gripping force.

In 2011 O'Doherty and colleagues showed a BCI with sensory feedback with rhesus monkeys. The monkey controlled the position of an avatar arm while receiving sensory feedback through direct intracortical stimulation (ICMS) in the arm representation area of the sensory cortex.

Donoghue, Schwartz, and Andersen

Other laboratories that have developed BCIs and algorithms that decode neuron signals include John Donoghue at the Carney Institute for Brain Science at Brown University, Andrew Schwartz at the University of Pittsburgh, and Richard Andersen at Caltech. These researchers produced working BCIs using recorded signals from far fewer neurons than Nicolelis (15–30 neurons versus 50–200 neurons).

The Carney Institute reported training rhesus monkeys to use a BCI to track visual targets on a computer screen (closed-loop BCI) with or without a joystick. The group created a BCI for three-dimensional tracking in virtual reality and reproduced BCI control in a robotic arm. The same group demonstrated that a monkey could feed itself pieces of fruit and marshmallows using a robotic arm controlled by the animal's brain signals.

Andersen's group used recordings of premovement activity from the posterior parietal cortex, including signals created when experimental animals anticipated receiving a reward.

Other Research

In addition to predicting kinematic and kinetic parameters of limb movements, BCIs that predict electromyographic or electrical activity of the muscles of primates are in process. Such BCIs could restore mobility in paralyzed limbs by electrically stimulating muscles.

Nicolelis and colleagues demonstrated that large neural ensembles can predict arm position. This work allowed BCIs to read arm movement intentions and translate them into actuator movements. Carmena and colleagues programmed a BCI that allowed a monkey to control reaching and grasping movements by a robotic arm. Lebedev and colleagues argued that brain networks reorganize to create a new representation of the robotic appendage in addition to the representation of the animal's own limbs.

In 2019, a study reported a BCI that had the potential to help patients with speech impairment caused by neurological disorders. Their BCI used high-density electrocorticography to tap neural activity from a patient's brain and used deep learning to synthesize speech. In 2021, those researchers reported the potential of a BCI to decode words and sentences in an anarthric patient who had been unable to speak for over 15 years.

The biggest impediment to BCI technology is the lack of a sensor modality that provides safe, accurate and robust access to brain signals. The use of a better sensor expands the range of communication functions that can be provided using a BCI.

Development and implementation of a BCI system is complex and time-consuming. In response to this problem, Gerwin Schalk has been developing BCI2000, a general-purpose system for BCI research, since 2000.

A new 'wireless' approach uses light-gated ion channels such as channelrhodopsin to control the activity of genetically defined subsets of neurons *in vivo*. In the context of a simple learning task, illumination of transfected cells in the somatosensory cortex influenced decision-making in mice.

BCIs led to a deeper understanding of neural networks and the central nervous system. Research has reported that despite neuroscientists' inclination to believe that neurons have the most effect when working together, single neurons can be conditioned through the use of BCIs to fire in a pattern that allows primates to control motor outputs. BCIs led to development of the single neuron insufficiency principle that states that even with a well-tuned firing rate, single neurons can only carry limited information and therefore the highest level of accuracy is achieved by recording ensemble firings. Other principles discovered with BCIs include the neuronal multitasking principle, the neuronal mass principle, the neural degeneracy principle, and the plasticity principle.

BCIs are proposed to be applied by users without disabilities. Passive BCIs allow for assessing and interpreting changes in the user state during Human-Computer Interaction (HCI). In a secondary, implicit control loop, the system adapts to its user, improving its usability.

BCI systems can potentially be used to encode signals from the periphery. These sensory BCI devices enable real-time, behaviorally-relevant decisions based upon closed-loop neural stimulation.

3.1.4 Human Research

Invasive BCIs

Invasive BCI requires surgery to implant electrodes under the scalp for accessing brain signals. The main advantage is to increase accuracy. Downsides include side effects from the surgery, including scar tissue that can obstruct brain signals or the body may not accept the implanted electrodes.

Vision

Invasive BCI research has targeted repairing damaged sight and providing new functionality for people with paralysis. Invasive BCIs are implanted directly into the grey matter of the brain during neurosurgery. Because they lie in the grey matter, invasive devices produce the highest quality signals of BCI devices but are prone to scar-tissue build-up, causing the signal to weaken, or disappear, as the body reacts to the foreign object.

In vision science, direct brain implants have been used to treat non-congenital (acquired) blindness. One of the first scientists to produce a working brain interface to restore sight was private researcher William Dobelle. Dobelle's first prototype was implanted into "Jerry", a man blinded in adulthood, in 1978. A single-array BCI containing 68 electrodes was implanted onto Jerry's visual cortex and succeeded in producing phosphenes, the sensation of seeing light. The system included cameras mounted on glasses to send signals to the implant. Initially, the implant allowed Jerry to see shades of grey in a limited field of vision at a low frame-rate. This also required him to be hooked up to a mainframe computer, but shrinking electronics and faster computers made his artificial eye more portable and now enable him to perform simple tasks unassisted.

In 2002, Jens Naumann, also blinded in adulthood, became the first in a series of 16 paying patients to receive Dobelle's second generation implant, one of the earliest commercial uses of BCIs. The second generation device used a more sophisticated implant enabling better mapping of phosphenes into coherent vision. Phosphenes are spread out across the visual field in what researchers call "the starry-night effect". Immediately after his implant, Jens was able to use his imperfectly restored vision to drive an automobile slowly around the parking area of the research institute. Dobelle died in 2004 before his processes and developments were documented, leaving no one to continue his work. Subsequently, Naumann and the other patients in the program began having problems with their vision, and eventually lost their "sight" again.

Movement

BCIs focusing on motor neuroprosthetics aim to restore movement in individuals with paralysis or provide devices to assist them, such as interfaces with computers or robot arms. Kennedy and Bakay were first to install a human brain implant that produced signals of high enough quality to simulate movement. Their patient, Johnny Ray (1944–2002), developed 'locked-in syndrome' after a brain-stem stroke in 1997. Ray's implant was installed in 1998 and he lived long enough to start working with the implant, eventually learning to control a computer cursor; he died in 2002 of a brain aneurysm.

Tetraplegic Matt Nagle became the first person to control an artificial hand using a BCI in 2005 as part of the first nine-month human trial of Cyberkinetics's BrainGate chip-implant. Implanted in Nagle's right precentral gyrus (area of the motor cortex for arm movement), the 96-electrode implant allowed Nagle to control a robotic arm by thinking about moving his hand as well as a computer cursor, lights and TV. One year later, Jonathan Wolpaw received the Altran Foundation for Innovation prize for developing a Brain Computer Interface with electrodes located on the surface of the skull, instead of directly in the brain.

Research teams led by the BrainGate group and another at University of Pittsburgh Medical Center, both in collaborations with the United States Department of Veterans Affairs (VA), demonstrated control of prosthetic limbs with many degrees of freedom using direct connections to arrays of neurons in the motor cortex of tetraplegia patients.

Communication

In May 2021, a Stanford University team reported a successful proof-of-concept test that enabled a quadraplegic participant to produce English sentences at about 86 characters per minute and 18 words per minute. The participant imagined moving his hand to write letters, and the system performed handwriting recognition on electrical signals detected in the motor cortex, utilizing Hidden Markov models and recurrent neural networks.

A 2021 study reported that a paralyzed patient was able to communicate 15 words per minute using a brain implant that analyzed vocal tract motor neurons.

In a review article, authors wondered whether human information transfer rates can surpass that of language with BCIs. Language research has reported that information transfer rates are relatively constant across many languages. This may reflect the brain's information processing limit. Alternatively, this limit may be intrinsic to language itself, as a modality for information transfer.

In 2023 two studies used BCIs with recurrent neural network to decode speech at a record rate of 62 words per minute and 78 words per minute.

Technical Challenges

There exist a number of technical challenges to recording brain activity with invasive BCIs. Advances in CMOS technology are pushing and enabling integrated, invasive BCI designs with smaller size, lower power requirements, and higher signal acquisition capabilities. Invasive BCIs involve electrodes that penetrate brain tissue in an attempt to record action potential signals (also known as spikes) from individual, or small groups of, neurons near the electrode. The interface between a recording electrode and the electrolytic solution surrounding neurons has been modelled using the Hodgkin-Huxley model.

Electronic limitations to invasive BCIs have been an active area of research in recent decades. While intracellular recordings of neurons reveal action potential voltages on the scale of hundreds of millivolts, chronic invasive BCIs rely on recording extracellular voltages which typically are three orders of magnitude smaller, existing at hundreds of microvolts. Further adding to the challenge of detecting signals on the scale of microvolts is the fact that the electrode-tissue interface has a high capacitance at small voltages. Due to the nature of these small signals, for BCI systems that incorporate functionality onto an integrated circuit, each electrode requires its own amplifier and ADC, which convert analog extracellular voltages into digital signals. Because a typical neuron action potential lasts for one millisecond, BCIs measuring spikes must have sampling rates ranging from 300 Hz to 5 kHz. Yet another concern is that invasive BCIs must be low-power, so as to dissipate less heat to surrounding tissue; at the most basic level more power is traditionally needed to optimize signal-to-noise ratio. Optimal battery design is an active area of research in BCIs.

Challenges existing in the area of material science are central to the design of invasive BCIs. Variations in signal quality over time have been commonly observed with implantable microelectrodes. Optimal material and mechanical characteristics for long term signal stability in invasive BCIs has been an active area of research. It has been proposed that the formation of glial scarring, secondary to damage at the electrode-tissue interface, is likely responsible for electrode failure and reduced recording performance. Research has suggested that blood-brain barrier leakage, either at the time of insertion or over time, may be responsible for the inflammatory and glial reaction to chronic microelectrodes implanted in the brain. As a result, flexible and tissue-like designs have been researched and developed to minimize foreign-body reaction by means of matching the Young's modulus of the electrode closer to that of brain tissue.

Partially Invasive BCIs

Partially invasive BCI devices are implanted inside the skull but rest outside the brain rather than within the grey matter. They produce higher resolution signals than non-invasive BCIs where the bone tissue of the cranium deflects and deforms signals and have a lower risk of forming scar-tissue in the brain than fully invasive

BCIs. Preclinical demonstration of intracortical BCIs from the stroke perilesional cortex has been conducted.

Endovascular

In 2010, researchers affiliated with University of Melbourne began developing a BCI that could be inserted via the vascular system. Australian neurologist Thomas Oxley conceived the idea for this BCI, called Stentrode, earning funding from DARPA. Preclinical studies evaluated the technology in sheep.

Stentrode is a monolithic stent electrode array, is designed to be delivered via an intravenous catheter under image-guidance to the superior sagittal sinus, in the region which lies adjacent to the motor cortex. This proximity enables Stentrode to measure neural activity. The procedure is most similar to how venous sinus stents are placed for the treatment of idiopathic intracranial hypertension. Stentrode communicates neural activity to a battery-less telemetry unit implanted in the chest, which communicates wirelessly with an external telemetry unit capable of power and data transfer. While an endovascular BCI benefits from avoiding a craniotomy for insertion, risks such as clotting and venous thrombosis exist.

Human trials with Stentrode were underway as of 2021. In November 2020, two participants with amyotrophic lateral sclerosis were able to wirelessly control an operating system to text, email, shop, and bank using direct thought using Stentrode, marking the first time a brain-computer interface was implanted via the patient's blood vessels, eliminating the need for brain surgery. In January 2023, researchers reported no serious adverse events during the first year for all four patients, who could use it to operate computers.

Electrocorticography

Electrocorticography (ECoG) measures brain electrical activity from beneath the skull in a way similar to non-invasive electroencephalography, using electrodes embedded in a thin plastic pad placed above the cortex, beneath the dura mater. ECoG technologies were first trialled in humans in 2004 by Eric Leuthardt and Daniel Moran from Washington University in St. Louis. In a later trial, the researchers enabled a teenage boy to play Space Invaders. This research indicates that control is rapid, requires minimal training, balancing signal fidelity and level of invasiveness.

Signals can be either subdural or epidural, but are not taken from within the brain parenchyma. Patients are required to have invasive monitoring for localization and resection of an epileptogenic focus.

ECoG offers higher spatial resolution, better signal-to-noise ratio, wider frequency range, and less training requirements than scalp-recorded EEG, and at the same time has lower technical difficulty, lower clinical risk, and may have superior long-term stability than intracortical single-neuron recording. This feature profile and evidence of the high level of control with minimal training requirements shows potential for real world application for people with motor disabilities.

Edward Chang and Joseph Makin from UCSF reported that ECoG signals could be used to decode speech from epilepsy patients implanted with high-density ECoG arrays over the peri-Sylvian cortices. They reported word error rates of 3% (a marked improvement from prior efforts) utilizing an encoder-decoder neural network, which translated ECoG data into one of fifty sentences composed of 250 unique words.

Non-invasive BCIs

Human experiments have used non-invasive neuroimaging interfaces. The majority of published BCI research involves noninvasive EEG-based BCIs. EEG-based technologies and interfaces have been used for the broadest variety of applications. Although EEG-based interfaces are easy to wear and do not require surgery, they have relatively poor spatial resolution and cannot effectively use higher-frequency signals because the skull interferes, dispersing and blurring the electromagnetic waves created by the neurons. EEG-based interfaces also require some time and effort prior to each usage session, while others require no prior-usage training. The choice of a specific BCI for a patient depends on numerous factors.

Functional Near-Infrared Spectroscopy

In 2014 and 2017, a BCI using functional near-infrared spectroscopy for "locked-in" patients with amyotrophic lateral sclerosis (ALS) was able to restore basic ability to communicate.

Electroencephalography (EEG)-Based Brain-Computer Interfaces

After Vidal stated the BCI challenge, the initial reports on non-invasive approaches included control of a cursor in 2D using VEP, control of a buzzer using CNV, control of a physical object, a robot, using a brain rhythm (alpha), control of a text written on a screen using P300.

In the early days of BCI research, another substantial barrier to using EEG was that extensive training ws required. For example, in experiments beginning in the mid-1990s, Niels Birbaumer at the University of Tübingen in Germany trained paralysed people to self-regulate the slow cortical potentials in their EEG to such an extent that these signals could be used as a binary signal to control a computer cursor. (Birbaumer had earlier trained epileptics to prevent impending fits by controlling this low voltage wave.)

The experiment trained ten patients to move a computer cursor. The process was slow, requiring more than an hour for patients to write 100 characters with the cursor, while training often took months. The slow cortical potential approach has fallen away in favor of approaches that require little or no training, are faster and more accurate, and work for a greater proportion of users.

Another research parameter is the type of oscillatory activity that is measured. Gert Pfurtscheller founded the BCI Lab 1991 and conducted the first online BCI based on oscillatory features and classifiers. Together with Birbaumer and Jonathan Wolpaw at New York State University they focused on developing technology that would allow users to choose the brain signals they found easiest to operate a BCI, including *mu* and *beta* rhythms.

A further parameter is the method of feedback used as shown in studies of P300 signals. Patterns of P300 waves are generated involuntarily (stimulus-feedback) when people see something they recognize and may allow BCIs to decode categories of thoughts without training.

A 2005 study reported EEG emulation of digital control circuits, using a CNV flip-flop. A 2009 study reported noninvasive EEG control of a robotic arm using a CNV flip-flop. A 2011 study reported control of two robotic arms solving Tower of Hanoi task with three disks using a CNV flip-flop. A 2015 study described EEG-emulation of a Schmitt trigger, flip-flop, demultiplexer, and modem.

Advances by Bin He and his team at University of Minnesota suggest the potential of EEG-based brain-computer interfaces to accomplish tasks close to invasive brain-computer interfaces. Using advanced functional neuroimaging including BOLD functional MRI and EEG source imaging, They identified the co-variation and co-localization of electrophysiological and hemodynamic signals. Refined by a neuroimaging approach and a training protocol, They fashioned a non-invasive EEG based brain-computer interface to control the flight of a virtual helicopter in 3-dimensional space, based upon motor imagination. In June 2013 they announced a technique to guide a remote-control helicopter through an obstacle course. They also solved the EEG inverse problem and then used the resulting virtual EEG for BCI tasks. Well-controlled studies suggested the merits of such a source analysis-based BCI.

A 2014 study reported that severely motor-impaired patients could communicate faster and more reliably with non-invasive EEG BCI than with muscle-based communication channels.

A 2019 study reported that the application of evolutionary algorithms could improve EEG mental state classification with a non-invasive Muse device, enabling classification of data acquired by a consumer-grade sensing device.

In a 2021 systematic review of randomized controlled trials using BCI for post-stroke upper-limb rehabilitation, EEG-based BCI was reported to have efficacy in improving upper-limb motor function compared to control therapies. More specifically, BCI studies that utilized band power features, motor imagery, and functional electrical stimulation were reported to be more effective than alternatives. Another 2021 systematic review focused on post-stroke robot-assisted EEG-based BCI for hand rehabilitation. Improvement in motor assessment scores was observed in three of eleven studies.

Dry Active Electrode Arrays

In the early 1990s Babak Taheri, at University of California, Davis demonstrated the first single and multichannel dry active electrode arrays. The arrayed electrode was demonstrated to perform well compared to silver/silver chloride electrodes. The device consisted of four sensor sites with integrated electronics to reduce noise by impedance matching. The advantages of such electrodes are:

- no electrolyte used,
- no skin preparation,
- significantly reduced sensor size, and
- compatibility with EEG monitoring systems.

The active electrode array is an integrated system containing an array of capacitive sensors with local integrated circuitry packaged with batteries to power the circuitry. This level of integration was required to achieve the result.

The electrode was tested on a test bench and on human subjects in four modalities, namely:

- spontaneous EEG,
- sensory event-related potentials,
- brain stem potentials, and
- cognitive event-related potentials.

Performance compared favorably with that of standard wet electrodes in terms of skin preparation, no gel requirements (dry), and higher signal-to-noise ratio.

In 1999 Hunter Peckham and others at Case Western Reserve University used a 64-electrode EEG skullcap to return limited hand movements to a quadriplegic. As he concentrated on simple but opposite concepts like up and down. A basic pattern was identified in his beta-rhythm EEG output and used to control a switch: Above average activity was interpreted as on, below average off. The signals were also used to drive nerve controllers embedded in his hands, restoring some movement.

SSVEP Mobile EEG BCIs

In 2009, the NCTU Brain-Computer-Interface-headband was reported. Those researchers also engineered silicon-based microelectro-mechanical system (MEMS) dry electrodes designed for application to non-hairy body sites. These electrodes were secured to the headband's DAQ board with snap-on electrode holders. The signal processing module measured alpha activity and transferred it over Bluetooth to a phone that assessed the patients' alertness and cognitive capacity. When the subject became drowsy, the phone sent arousing feedback to the operator to rouse them.

In 2011, researchers reported a cellular based BCI that could cause phone to ring. The wearable system was composed of a four channel bio-signal acquisition/ amplification module, a communication module, and a Bluetooth phone. The electrodes were placed to pick up steady state visual evoked potentials (SSVEPs). SSVEPs are electrical responses to flickering visual stimuli with repetition rates over 6 Hz that are best found in the parietal and occipital scalp regions of the visual cortex. It was reported that all study participants were able to initiate the phone call with minimal practice in natural environments.

The scientists reported that a single channel fast Fourier transform (FFT) and multiple channel system canonical correlation analysis (CCA) algorithm can support mobile BCIs. The CCA algorithm has been applied in experiments investigating BCIs with claimed high accuracy and speed. Cellular BCI technology can reportedly be translated for other applications, such as picking up sensorimotor mu/beta rhythms to function as a motor-imagery based BCI.

In 2013, comparative tests performed on Android cell phone, tablet, and computer based BCIs, analyzed the power spectrum density of resultant EEG SSVEPs. The stated goals of this study were to "increase the practicability, portability, and ubiquity of an SSVEP-based BCI, for daily use". It was reported that the stimulation frequency on all mediums was accurate, although the phone's signal was not stable. The amplitudes of the SSVEPs for the laptop and tablet were reported to be larger than those of the cell phone. These two qualitative characterizations were suggested as indicators of the feasibility of using a mobile stimulus BCI.

One of the difficulties with EEG readings is susceptibility to motion artifacts. In most research projects, the participants were asked to sit still in a laboratory setting, reducing head and eye movements as much as possible. However, since these initiatives were intended to create a mobile device for daily use, the technology had to be tested in motion. In 2013, researchers tested mobile EEG-based BCI technology, measuring SSVEPs from participants as they walked on a treadmill. Reported results were that as speed increased, SSVEP detectability using CCA decreased. Independent component analysis (ICA) had been shown to be efficient in separating EEG signals from noise. The researchers stated that CCA data with and without ICA processing were similar. They concluded that CCA demonstrated robustness to motion artifacts. EEG-based BCI applications offer low spatial resolution. Possible solutions include: EEG source connectivity based on graph theory, EEG pattern recognition based on Topomap and EEG-fMRI fusion.

Prosthesis and Environment Control

Non-invasive BCIs have been applied to prosthetic upper and lower extremity devices in people with paralysis. For example, Gert Pfurtscheller of Graz University of Technology and colleagues demonstrated a BCI-controlled functional electrical stimulation system to restore upper extremity movements in a person with tetraplegia due to spinal cord injury. Between 2012 and 2013, researchers at University of California, Irvine demonstrated for the first time that BCI technology can restore brain-controlled walking after spinal cord injury. In their study, a person with paraplegia operated a BCI-robotic gait orthosis to regain basic ambulation. In 2009 independent researcher Alex Blainey used the Emotiv EPOC to control a 5 axis robot arm. He made several demonstrations of mind controlled wheelchairs and home automation.

Magnetoencephalography and Fmri

Magnetoencephalography (MEG) and functional magnetic resonance imaging (fMRI) have both been used as non-invasive BCIs. In a widely reported experiment, fMRI allowed two users to play Pong in real-time by altering their haemodynamic response or brain blood flow through biofeedback.

fMRI measurements of haemodynamic responses in real time have also been used to control robot arms with a seven-second delay between thought and movement.

In 2008 research developed in the Advanced Telecommunications Research (ATR) Computational Neuroscience Laboratories in Kyoto, Japan, allowed researchers to reconstruct images from brain signals at a resolution of 10x10 pixels.

A 2011 study reported second-by-second reconstruction of videos watched by the study's subjects, from fMRI data. This was achieved by creating a statistical model relating videos to brain activity. This model was then used to look up 100 one-second video segments, in a database of 18 million seconds of random YouTube videos, matching visual patterns to brain activity recorded when subjects watched a video. These 100 one-second video extracts were then combined into a mash-up image that resembled the video.

BCI Control Strategies in Neurogaming

Motor Imagery:

- Motor imagery involves imagining the movement of body parts, activating the sensorimotor cortex, which modulates sensorimotor oscillations in the EEG. This can be detected by the BCI and used to infer user intent. Motor imagery typically requires training to acquire acceptable control. Training sessions typically consume hours over several days. Regardless of the duration of the training session, users are unable to master the control scheme. This results in very slow pace of the gameplay. Machine learning methods were used to compute a subject-specific model for detecting motor imagery performance. The top performing algorithm from BCI Competition IV in 2022 dataset 2 for motor imagery was the Filter Bank Common Spatial Pattern, developed by Ang et al. from A*STAR, Singapore.

Bio/Neurofeedback for Passive BCI Designs:

- Biofeedback can be used to monitor a subject's mental relaxation. In some cases, biofeedback does not match EEG, while parameters such as electromyography (EMG), galvanic skin resistance (GSR), and heart rate variability (HRV) can do so. Many biofeedback systems treat disorders such as attention deficit hyperactivity disorder (ADHD), sleep problems in children, teeth grinding, and chronic pain. EEG biofeedback systems typically monitor four brainwave bands (theta: 4–7 Hz, alpha:8–12 Hz, SMR: 12–15 Hz, beta: 15–18 Hz) and challenge the subject to control them. Passive BCI uses BCI to enrich human–machine interaction with information on the user's mental state, for example, simulations that detect when users intend to push brakes during emergency vehicle braking. Game developers using passive BCIs understand that through repetition of game levels the user's cognitive state adapts. During the first play of a given level, the player reacts differently than during subsequent plays: for example, the user is less surprised by an event that they expect.

Visual Evoked Potential (VEP):

- A VEP is an electrical potential recorded after a subject is presented with a visual stimuli. The types of VEPs include SSVEPs and P300 potential.
- Steady-state visually evoked potentials (SSVEPs) use potentials generated by exciting the retina, using visual stimuli modulated at certain frequencies. SSVEP stimuli are often formed from alternating checkerboard patterns and at times use flashing images. The frequency of the phase reversal of the stimulus used can be distinguished by EEG; this makes detection of SSVEP stimuli relatively easy. SSVEP is used within many BCI systems. This is due to several factors. The signal elicited is measurable in as large a population as the transient VEP and blink movement. Electrocardiographic artefacts do not affect the frequencies monitored. The SSVEP signal is robust; the topographic

organization of the primary visual cortex is such that a broader area obtains afferents from the visual field's central or fovial region. SSVEP comes with problems. As SSVEPs use flashing stimuli to infer user intent, the user must gaze at one of the flashing or iterating symbols in order to interact with the system. It is, therefore, likely that the symbols become irritating and uncomfortable during longer play sessions.

- Another type of VEP is the P300 potential. This potential is a positive peak in the EEG that occurs roughly 300 ms after the appearance of a target stimulus (a stimulus for which the user is waiting or seeking) or oddball stimuli. P300 amplitude decreases as the target stimuli and the ignored stimuli grow more similar. P300 is thought to be related to a higher level attention process or an orienting response. Using P300 requires fewer training sessions. The first application to use it was the P300 matrix. Within this system, a subject chooses a letter from a 6 by 6 grid of letters and numbers. The rows and columns of the grid flashed sequentially and every time the selected "choice letter" was illuminated the user's P300 was (potentially) elicited. However, the communication process, at approximately 17 characters per minute, was slow. P300 offers a discrete selection rather than continuous control. The advantage of P300 within games is that the player does not have to learn how to use a new control system, requiring only short training instances to learn gameplay mechanics and the basic BCI paradigm.

Non-Brain-Based Human–Computer Interface (Physiological Computing)

- Human-computer interaction can exploit other recording modalities, such as electrooculography and eye-tracking. These modalities do not record brain activity and therefore do not qualify as BCIs.

Electrooculography (EOG):

- In 1989, a study reported control of a mobile robot by eye movement using electrooculography signals. A mobile robot was driven to a goal point using five EOG commands, interpreted as forward, backward, left, right, and stop.

Pupil-size oscillation:

- A new non-EEG-based HCI that required no visual fixation, or ability to move the eyes. The interface is based on covert interest; directing attention to a chosen letter on a virtual keyboard, without the need to look directly at the letter. Each letter has its own (background) circle which micro-oscillates in brightness differently from the others. Letter selection is based on best fit between unintentional pupil-size oscillation and the background circle's brightness oscillation pattern. Accuracy is additionally improved by the user's mental rehearsal of the words 'bright' and 'dark' in synchrony with the brightness transitions of the letter's circle.

Brain-to-Brain Communication

In the 1960s a researcher after training used EEG to create Morse code using alpha waves. On 27 February 2013 Miguel Nicolelis's group at Duke University and IINN-ELS connected the brains of two rats, allowing them to share information, in the first-ever direct brain-to-brain interface.

Gerwin Schalk reported that ECoG signals can discriminate vowels and consonants embedded in spoken and imagined words, shedding light on the mechanisms associated with their production and could provide a basis for brain-based communication using imagined speech.

In 2002 Kevin Warwick had an array of 100 electrodes fired into his nervous system in order to link his nervous system to the Internet. Warwick carried out a series of experiments. Electrodes were implanted into his wife's nervous system, allowing them to conduct the first direct electronic communication experiment between the nervous systems of two humans. Other researchers achieved brain-to-brain communication between at a distance using non-invasive technology attached to the participants' scalps. The words were encoded in binary streams by the cognitive motor input of the person sending the information. Pseudo-random bits of the information carried encoded words "hola" ("hi" in Spanish) and "ciao" ("goodbye" in Italian) and were transmitted mind-to-mind.

3.1.5 Cell-Culture BCIs

Researchers have built devices to interface with neural cells and entire neural networks *in vitro*. Experiments on cultured neural tissue focused on building problem-solving networks, constructing basic computers and manipulating robotic devices. Research into techniques for stimulating and recording individual neurons grown on semiconductor chips is neuroelectronics or neurochips.

Development of the first neurochip was claimed by a Caltech team led by Jerome Pine and Michael Maher in 1997. The Caltech chip had room for 16 neurons.

In 2003 a team led by Theodore Berger, at the University of Southern California, worked on a neurochip designed to function as an artificial or prosthetic hippocampus. The neurochip was designed for rat brains. The hippocampus was chosen because it is thought to be the most structured and most studied part of the brain. Its function is to encode experiences for storage as long-term memories elsewhere in the brain.

In 2004 Thomas DeMarse at the University of Florida used a culture of 25,000 neurons taken from a rat's brain to fly a F-22 fighter jet aircraft simulator. After collection, the cortical neurons were cultured in a petri dish and reconnected themselves to form a living neural network. The cells were arranged over a grid of 60 electrodes and used to control the pitch and yaw functions of the simulator. The study's focus was on understanding how the human brain performs and learns computational tasks at a cellular level.

3.1.6 Collaborative BCIs

The idea of combining/integrating brain signals from multiple individuals was introduced at Humanity+ @Caltech, in December 2010, by Adrian Stoica, who referred to the concept as multi-brain aggregation. A patent was applied for in 2012. Stoica's first paper on the topic appeared in 2012, after the publication of his patent application.

3.1.7 Ethical Considerations

BCIs present significant ethical questions, including concerns about privacy, autonomy, consent, and the consequences of merging human cognition with external devices. Exploring these ethical considerations highlights the complex interplay between advancing technology and preserving fundamental human rights and values. The concerns can be broadly categorized into user-centric issues and legal and social issues.

Concerns center on the safety and long-term effects on users. These include obtaining informed consent from individuals with communication difficulties, the impact on patients' and families' quality of life, health-related side effects, misuse of therapeutic applications, safety risks, and the non-reversible nature of some BCI-induced changes. Additionally, questions arise about access to maintenance, repair, and spare parts, particularly in the event of a company's bankruptcy

The legal and social aspects of BCIs complicate mainstream adoption. Concerns include issues of accountability and responsibility, such as claims that BCI influence overrides free will and control over actions, inaccurate translation of cognitive intentions, personality changes resulting from deep-brain stimulation, and the blurring of the line between human and machine. Other concerns involve the use of BCIs in advanced interrogation techniques, unauthorized access ("brain hacking"), social stratification through selective enhancement, privacy issues related to mind-reading, tracking and "tagging" systems, and the potential for mind, movement, and emotion control. Researchers have also theorized that BCIs could exacerbate existing social inequalities.

In their current form, most BCIs are more akin to corrective therapies that engage few of such ethical issues. Bioethics is well-equipped to address the challenges posed by BCI technologies, with Clausen suggesting in 2009 that "BCIs pose ethical challenges, but these are conceptually similar to those that bioethicists have addressed for other realms of therapy." Haselager and colleagues highlighted the importance of managing expectations and value. Standard protocols can ensure ethically sound informed-consent procedures for locked-in patients.

The evolution of BCIs mirrors that of pharmaceutical science, which began as a means to address impairments and now enhances focus and reduces the need for sleep. As BCIs progress from therapies to enhancements, the BCI community is working to create consensus on ethical guidelines for research, development, and dissemination. Ensuring equitable access to BCIs will be crucial in preventing generational inequalities that could hinder the right to human flourishing.

3.1.8 Low-Cost Systems

Various companies are developing inexpensive BCIs for research and entertainment. Toys such as the NeuroSky and Mattel MindFlex have seen some commercial success.

- **In 2006:** Sony patented a neural interface system allowing radio waves to affect signals in the neural cortex.
- **In 2007:** NeuroSky released the first affordable consumer based EEG along with the game NeuroBoy. It was the first large scale EEG device to use dry sensor technology.
- **In 2008:** OCZ Technology developed a device for use in video games relying primarily on electromyography.
- **In 2008:** Final Fantasy developer Square Enix announced that it was partnering with NeuroSky to create Judecca, a game.
- **In 2009:** Mattel partnered with NeuroSky to release Mindflex, a game that used an EEG to steer a ball through an obstacle course. It was by far the best selling consumer based EEG at the time.
- **In 2009:** Uncle Milton Industries partnered with NeuroSky to release the Star Wars Force Trainer, a game designed to create the illusion of possessing the Force.
- **In 2009:** Emotiv released the EPOC, a 14 channel EEG device that can read 4 mental states, 13 conscious states, facial expressions, and head movements. The EPOC was the first commercial BCI to use dry sensor technology, which can be dampened with a saline solution for a better connection.
- **In November 2011:** *Time* magazine selected "necomimi" produced by Neurowear as one of the year's best inventions.

- **In February 2014:** They Shall Walk (a nonprofit organization fixed on constructing exoskeletons, dubbed LIFESUITs, for paraplegics and quadriplegics) began a partnership with James W. Shakarji on the development of a wireless BCI.
- **In 2016:** A group of hobbyists developed an open-source BCI board that sends neural signals to the audio jack of a smartphone, dropping the cost of entry-level BCI to £20. Basic diagnostic software is available for Android devices, as well as a text entry app for Unity.
- **In 2020:** NextMind released a dev kit including an EEG headset with dry electrodes at $399. The device can run various visual-BCI demonstration applications or developers can create their own. It was later acquired by Snap Inc. in 2022.

3.1.9 Future Directions

A consortium of 12 European partners completed a roadmap to support the European Commission in their funding decisions for the Horizon 2020 framework program. The project was funded by the European Commission. It started in November 2013 and published a roadmap in April 2015. A 2015 publication describes this project, as well as the Brain-Computer Interface Society. It reviewed work within this project that further defined BCIs and applications, explored recent trends, discussed ethical issues, and evaluated directions for new BCIs.

Other recent publications too have explored future BCI directions for new groups of disabled users.

Disorders of Consciousness (DOC)

Some people have a disorder of consciousness (DOC). This state is defined to include people in a coma and those in a vegetative state (VS) or minimally conscious state (MCS). BCI research seeks to address DOC.

A key initial goal is to identify patients who can perform basic cognitive tasks, which would change their diagnosis, and allow them to make important decisions (such as whether to seek therapy, where to live, and their views on end-of-life decisions regarding them). Patients incorrectly diagnosed may die as a result of end-of-life decisions made by others. The prospect of using BCI to communicate with such patients is a tantalizing prospect.

Many such patients cannot use BCIs based on vision. Hence, tools must rely on auditory and/or vibrotactile stimuli. Patients may wear headphones and/or vibrotactile stimulators placed on responsive body parts. Another challenge is that patients may be able to communicate only at unpredictable intervals. Home devices can allow communications when the patient is ready.

Automated tools can ask questions that patients can easily answer, such as "Is your father named George?" or "Were you born in the USA?" Automated instructions inform patients how to coney yes or no, for example by focusing their attention on stimuli on the right vs. left wrist. This focused attention produces reliable changes in EEG patterns that can help determine whether the patient is able to communicate.

Motor Recovery

People may lose some of their ability to move due to many causes, such as stroke or injury. Research in recent years has demonstrated the utility of EEG-based BCI systems in aiding motor recovery and neurorehabilitation in patients who have had a stroke. Several groups have explored systems and methods for motor recovery that include BCIs. In this approach, a BCI measures motor activity while the patient imagines or attempts movements as directed by a therapist. The BCI may provide two benefits:

- if the BCI indicates that a patient is not imagining a movement correctly (non-compliance), then the BCI could inform the patient and therapist; and
- rewarding feedback such as functional stimulation or the movement of a virtual avatar also depends on the patient's correct movement imagery.

So far, BCIs for motor recovery have relied on the EEG to measure the patient's motor imagery. However, studies have also used fMRI to study different changes in the brain as persons undergo BCI-based stroke rehab training. Imaging studies combined with EEG-based BCI systems hold promise for investigating neuroplasticity during motor recovery post-stroke. Future systems might include the fMRI and other measures for real-time control, such as functional near-infrared, probably in tandem with EEGs. Non-invasive brain stimulation has also been explored in combination with BCIs for motor recovery. In 2016, scientists out of the University of Melbourne published preclinical proof-of-concept data related to a potential brain-computer interface technology platform being developed for patients with paralysis to facilitate control of external devices such as robotic limbs, computers and exoskeletons by translating brain activity.

Functional Brain Mapping

In 2014, some 400,000 people underwent brain mapping during neurosurgery. This procedure is often required for people who do not respond to medication. During this procedure, electrodes are placed on the brain to precisely identify the locations of structures and functional areas. Patients may be awake during neurosurgery and asked to perform tasks, such as moving fingers or repeating words. This is necessary so that surgeons can remove the desired tissue while sparing other regions. Removing too much brain tissue can cause permanent damage, while removing too little can mandate additional neurosurgery.

Researchers explored ways to improve neurosurgical mapping. This work focuses largely on high gamma activity, which is difficult to detect non-invasively. Results improved methods for identifying key functional areas.

Flexible Devices

Flexible electronics are polymers or other flexible materials (e.g. silk, pentacene, PDMS, Parylene, polyimide) printed with circuitry; the flexibility allows the electronics to bend. The fabrication techniques used to create these devices resembles those used to create integrated circuits and microelectromechanical systems (MEMS). Flexible neural interfaces may minimize brain tissue trauma related to mechanical mismatch between electrode and tissue.

Neural Dust

Neural dust is millimeter-sized devices operated as wirelessly powered nerve sensors that were proposed in a 2011 paper from the University of California, Berkeley Wireless Research Center. In one model, local field potentials could be distinguished from action potential "spikes", which would offer greatly diversified data vs conventional techniques.

3.2 HAPTIC FEEDBACK AND TOUCH INTERFACES

Haptic feedback and touch interfaces are crucial components of modern human-computer interaction, providing users with tactile sensations that enhance the usability and experience of digital devices. These technologies bridge the gap between the virtual and physical worlds, offering more intuitive and immersive ways to interact with electronic systems.

Haptic feedback and touch interfaces represent a significant advancement in human-computer interaction, providing more intuitive, immersive, and satisfying ways to interact with digital devices. These technologies enhance user experiences, improve accessibility, and increase engagement across a wide range of applications, from consumer electronics and gaming to healthcare and automotive industries.

Despite the challenges of technical limitations, cost, standardization, and user adaptation, ongoing advances in haptic technology and the integration of AI promise to drive further innovation and expansion.

By addressing ethical considerations and ensuring equitable access, we can harness the full potential of haptic feedback and touch interfaces to enhance human interaction with the digital world.

3.2.1 Principles of Haptic Feedback

Definition and Mechanism

Haptic feedback, also known as tactile feedback, refers to the use of touch sensations to communicate information to the user. This feedback can be delivered through vibrations, forces, or motions that mimic the sense of touch. The mechanism of haptic feedback involves actuators that produce physical sensations in response to user interactions, providing real-time tactile responses that enhance the user's perception and control.

Types of Haptic Feedback

There are several types of haptic feedback:

- **Vibrotactile Feedback:** This is the most common form, where small vibrations are used to simulate touch sensations. It is widely used in mobile devices, gaming controllers, and wearable technology.
- **Force Feedback:** This involves the application of forces to the user's body to simulate the sensation of weight or resistance. It is commonly used in virtual reality (VR) systems and robotic surgery to provide a realistic sense of interaction with virtual objects.
- **Electrotactile Feedback:** This type uses electrical stimulation to activate nerve endings in the skin, creating a sensation of touch. It is less common but offers precise control over the tactile sensations.

Sensory Integration

Haptic feedback works by integrating with the user's sensory system, particularly the mechanoreceptors in the skin, which detect pressure, vibration, and texture. When combined with visual and auditory feedback, haptic feedback creates a multi-sensory experience that can significantly enhance the realism and immersion of digital interactions.

3.2.2 Development of Touch Interfaces

Historical Evolution

The development of touch interfaces has evolved significantly since their inception. Early touch screens, such as those used in ATMs and public information kiosks, relied on resistive technology, which required users to apply pressure to register input. Capacitive touch screens, introduced later, detect changes in electrical fields caused by the touch of a finger, offering more sensitive and responsive interfaces. The introduction of multi-touch technology, popularized by devices like the iPhone, enabled more complex gestures and interactions, revolutionizing the way users interact with digital devices.

Modern Touch Interface Technologies

Modern touch interfaces utilize a range of technologies to provide precise and responsive interactions:

- **Capacitive Touch Screens:** These detect touch by measuring changes in electrical capacitance. They are highly sensitive and support multi-touch gestures, making them ideal for smartphones, tablets, and touch-enabled laptops.
- **Infrared Touch Screens:** These use an array of infrared light beams across the screen surface. When a finger or stylus breaks the beams, the touch location is detected. This technology is often used in large interactive displays.
- **Surface Acoustic Wave (SAW) Touch Screens:** These screens use ultrasonic waves that pass over the screen surface. A touch is detected when the waves are absorbed by a finger or stylus, allowing precise input detection.
- **Haptic Touch Screens:** These incorporate haptic feedback mechanisms to provide tactile responses to touch interactions, enhancing the user's experience by simulating the feel of pressing physical buttons or interacting with textured surfaces.

3.2.3 Applications of Haptic Feedback and Touch Interfaces

Consumer Electronics

Haptic feedback and touch interfaces are ubiquitous in consumer electronics. Smartphones, tablets, and wearable devices utilize these technologies to provide intuitive and responsive user experiences. Haptic feedback enhances interactions by providing tactile confirmations for touch inputs, improving usability and satisfaction.

Virtual Reality and Gaming

In VR and gaming, haptic feedback plays a critical role in creating immersive experiences. VR gloves and controllers with haptic feedback allow users to feel virtual objects and environments, enhancing the sense of presence and realism. Gaming controllers with force feedback provide tactile sensations that simulate in-game actions, such as the recoil of a gun or the rumble of a car engine, increasing engagement and enjoyment.

Medical and Rehabilitation

Haptic feedback and touch interfaces are increasingly used in medical and rehabilitation settings. Surgical simulators with haptic feedback allow surgeons to practice procedures with realistic touch sensations, improving training and skill development. In rehabilitation, haptic devices can help patients regain motor skills by providing tactile feedback during exercises, enhancing the effectiveness of therapy.

Automotive Industry

In the automotive industry, touch interfaces and haptic feedback are used to improve driver interaction with vehicle controls. Touch screens with haptic feedback provide tactile responses that help drivers keep their eyes on the road while interacting with infotainment systems. Haptic feedback in steering wheels and pedals can also alert drivers to potential hazards, enhancing safety.

3.2.4 Benefits of Haptic Feedback and Touch Interfaces

Enhanced User Experience

Haptic feedback significantly enhances the user experience by providing tactile confirmations of interactions. This reduces errors and increases confidence in using digital devices, leading to a more satisfying and intuitive user experience.

Improved Accessibility

Touch interfaces with haptic feedback can improve accessibility for individuals with visual impairments. Tactile responses help users navigate touch screens and interact with digital content without relying solely on visual cues, making technology more inclusive.

Increased Engagement and Immersion

In applications such as gaming and VR, haptic feedback increases engagement and immersion by providing realistic touch sensations. This enhances the sense of presence and makes virtual experiences more compelling and enjoyable.

Precision and Control

Haptic feedback provides precise control in applications requiring fine motor skills, such as medical simulations and remote surgeries. Tactile sensations help users perform delicate tasks with greater accuracy and confidence.

3.2.5 Challenges of Haptic Feedback and Touch Interfaces

Technical Limitations

Technical limitations, such as the resolution and latency of haptic feedback, can affect the realism and effectiveness of touch interfaces. Developing more advanced haptic technologies that provide high-resolution, low-latency feedback is essential for improving user experiences.

Cost and Complexity

The cost and complexity of integrating haptic feedback into devices can be a barrier to widespread adoption. Advanced haptic systems require sophisticated hardware and software, increasing production costs and complexity. Finding cost-effective solutions is crucial for broader implementation.

Standardization and Compatibility

Lack of standardization and compatibility across different devices and platforms can hinder the development and adoption of haptic feedback technologies. Establishing industry standards and ensuring cross-platform compatibility are essential for fostering innovation and widespread use.

User Adaptation

Users may need time to adapt to new touch interfaces and haptic feedback systems. Ensuring that these technologies are intuitive and easy to use is crucial for user acceptance and satisfaction. Providing adequate training and support can help users adapt to new interfaces more quickly.

3.2.6 Future Directions

Advances in Haptic Technology

Advances in haptic technology will continue to drive the development of more realistic and immersive touch interfaces. Emerging technologies, such as electroactive polymers and microfluidic actuators, promise to provide more nuanced and dynamic haptic feedback, enhancing the range and realism of tactile sensations.

Integration with AI and Machine Learning

The integration of artificial intelligence (AI) and machine learning with haptic feedback systems will enable more personalized and adaptive user experiences. AI can analyze user interactions and optimize haptic responses in real-time, improving the effectiveness and satisfaction of touch interfaces.

Expanding Applications

The applications of haptic feedback and touch interfaces will continue to expand into new areas, such as education, remote work, and smart home technologies. Haptic feedback can enhance remote collaboration by providing tactile sensations that simulate physical presence, while touch interfaces can make smart home devices more intuitive and accessible.

Ethical Considerations

As haptic feedback and touch interfaces become more pervasive, ethical considerations must be addressed. Ensuring privacy and data security, especially in medical and personal applications, is critical. Additionally, considerations around the equitable access to advanced haptic technologies and their potential impact on job displacement and societal norms should be carefully examined.

3.3 AUGMENTATION AND VIRTUAL REALITY IN HMIS

Human-Machine Interfaces (HMIs) are evolving rapidly, driven by advancements in augmentation and virtual reality (VR) technologies. These innovations are transforming how humans interact with machines, creating more immersive, intuitive, and efficient experiences.

Augmentation and virtual reality are revolutionizing human-machine interfaces, creating more immersive, intuitive, and efficient ways for humans to interact with machines.

These technologies have diverse applications across healthcare, education, industry, gaming, and retail, offering significant benefits such as enhanced user experience, improved training, increased efficiency, and enhanced collaboration. However, challenges such as technical limitations, cost, user adaptation, and privacy concerns must be addressed to fully realize the potential of AR and VR in HMIs.

Future advancements in technology and integration with AI, along with careful consideration of ethical issues, will shape the future of augmentation and VR, transforming how we interact with the digital world and enhancing our capabilities.

3.3.1 Principles of Augmentation and Virtual Reality

Augmentation

Augmentation refers to enhancing human capabilities through technology. In the context of HMIs, augmentation typically involves overlaying digital information onto the physical world, enhancing the user's perception and interaction with their environment. Augmented Reality (AR) is a prominent example, where digital elements are superimposed onto the real world via devices like smartphones, tablets, or AR glasses. This creates a blended experience that enriches real-world interactions with virtual data.

Virtual Reality

Virtual Reality (VR) involves creating entirely immersive digital environments that users can interact with in a seemingly real or physical way. VR typically requires specialized equipment such as VR headsets, gloves, and motion controllers to fully engage the user's senses. The aim is to transport the user to a different reality, offering a fully immersive experience that can be used for gaming, training, simulations, and more.

Human-Machine Interfaces

HMIs are the mediums through which humans interact with machines. Traditional HMIs include keyboards, mice, and touch screens. However, with the advent of AR and VR, HMIs are becoming more sophisticated, allowing for more natural and intuitive interactions. These interfaces are now incorporating voice commands, gestures, and even neural inputs, creating seamless communication between humans and machines.

3.3.2 Applications of Augmentation and VR in HMIs

Healthcare

In healthcare, AR and VR are revolutionizing medical training, diagnostics, and treatment. AR can overlay critical information during surgeries, providing real-time data and visual guidance to surgeons. VR is used for surgical simulations, allowing medical professionals to practice complex procedures in a risk-free environment. Additionally, VR is employed in rehabilitation, helping patients recover motor skills through immersive exercises.

Education and Training

AR and VR are transforming education and training by creating interactive and engaging learning experiences. AR can bring textbooks to life by overlaying 3D models and interactive elements on printed material, enhancing comprehension and retention. VR offers immersive simulations for training purposes, such as virtual laboratories for science students or flight simulators for pilots, providing hands-on experience without the associated risks.

Industrial and Manufacturing

In the industrial and manufacturing sectors, AR and VR improve efficiency and safety. AR can guide workers through complex assembly processes with step-by-step instructions overlaid on physical components. VR is used for training employees in hazardous environments, allowing them to practice procedures in a controlled, virtual setting. These technologies also facilitate remote collaboration, enabling experts to assist on-site workers from afar.

Gaming and Entertainment

The gaming and entertainment industry has been one of the earliest adopters of AR and VR. These technologies offer immersive gaming experiences, transporting players into virtual worlds where they can interact with the environment and other players in real-time. AR games, like Pokémon GO, blend the physical and digital worlds, creating interactive experiences in everyday settings. VR movies and experiences provide new ways to consume media, offering 360-degree views and interactive elements.

Retail and Marketing

AR and VR are enhancing the retail and marketing sectors by providing immersive shopping experiences. AR allows customers to visualize products in their real environment before making a purchase, such as trying on virtual clothes or placing virtual furniture in their home. VR offers virtual stores where customers can browse and shop in an immersive, interactive environment, improving customer engagement and satisfaction.

3.3.3 Benefits of Augmentation and VR in HMIs

Enhanced User Experience

One of the primary benefits of AR and VR in HMIs is the enhanced user experience. These technologies provide more intuitive and immersive interactions, making complex tasks easier and more enjoyable. Users can interact with digital content in a natural way, using gestures, voice commands, and even eye movements, creating a seamless and engaging experience.

Improved Training and Education

AR and VR significantly improve training and education by providing hands-on, interactive experiences. These technologies allow learners to practice skills in a safe and controlled environment, reducing the risk of errors and enhancing retention. AR and VR also offer personalized learning experiences, adapting to the learner's pace and style, improving outcomes.

Increased Efficiency and Productivity

In industrial and manufacturing settings, AR and VR can increase efficiency and productivity by providing real-time guidance and reducing the need for physical prototypes. Workers can access detailed instructions and information without interrupting their workflow, and virtual simulations can identify potential issues before they arise, saving time and resources.

Enhanced Collaboration

AR and VR facilitate enhanced collaboration by enabling remote interactions in a shared virtual space. Teams can work together on projects, conduct meetings, and solve problems without being physically present in the same location. This is particularly beneficial in global industries, where expertise and resources can be pooled from different parts of the world.

3.3.4 Challenges of Augmentation and VR in HMIs

Technical Limitations

Despite the advancements, there are still technical limitations to AR and VR technologies. Issues such as latency, limited field of view, and the resolution of displays can affect the user experience. Additionally, the development of high-quality AR and VR content requires significant computational power and expertise, which can be costly and time-consuming.

Cost and Accessibility

The cost of AR and VR hardware and software can be prohibitive, limiting accessibility for many users and organizations. High-end VR headsets and AR glasses are expensive, and developing custom applications can incur substantial costs. Ensuring broader accessibility to these technologies is crucial for widespread adoption and utilization.

User Adaptation and Acceptance

Users may face challenges in adapting to and accepting AR and VR technologies. Some individuals may experience discomfort or motion sickness when using VR headsets, and there may be a learning curve associated with new interfaces. Educating users and providing adequate support and training can help mitigate these issues and improve acceptance.

Privacy and Security

The use of AR and VR in HMIs raises concerns about privacy and security. These technologies often require access to personal data and real-world environments, which can pose risks if not properly managed. Ensuring robust data protection measures and addressing privacy concerns are essential to building trust and safeguarding users.

3.3.5 Future Directions

Advances in Technology

Future advancements in AR and VR technology will focus on improving resolution, reducing latency, and enhancing the realism of virtual experiences. Innovations in display technology, such as microLEDs and holographic displays, will provide higher quality visuals, while advancements in haptic feedback will offer more realistic tactile sensations. Improved motion tracking and neural interfaces will create more seamless and intuitive interactions.

Integration with AI

The integration of artificial intelligence (AI) with AR and VR will unlock new possibilities for HMIs. AI can enhance the functionality of these technologies by providing real-time data analysis, personalized experiences, and intelligent virtual assistants. Machine learning algorithms can improve object recognition, natural language processing, and predictive analytics, making AR and VR more powerful and versatile.

Expanding Applications

As AR and VR technologies continue to evolve, their applications will expand into new areas. In healthcare, AR and VR will enable more advanced diagnostics, telemedicine, and personalized treatments. In education, these technologies will offer more immersive and interactive learning experiences, catering to diverse learning styles and needs. In everyday life, AR and VR will enhance smart home systems, providing intuitive and interactive interfaces for managing household tasks and entertainment.

Ethical Considerations

The widespread adoption of AR and VR in HMIs will necessitate careful consideration of ethical issues. Ensuring equitable access to these technologies, addressing privacy and security concerns, and preventing potential misuse are critical. Establishing ethical guidelines and regulations will help ensure that AR and VR are developed and used responsibly, benefiting society as a whole.

CHAPTER 4
BIONICS AND PROSTHETICS

4.1 ADVANCEMENTS IN PROSTHETIC LIMBS

Prosthetic limbs have come a long way from their rudimentary beginnings, evolving into highly sophisticated devices that can significantly enhance the quality of life for individuals with limb loss.

Recent advancements in materials science, robotics, neurology, and biomechanics have led to the development of prosthetics that are not only functional but also more comfortable, intuitive, and integrated with the human body.

Advancements in prosthetic limbs have transformed the lives of individuals with limb loss, offering enhanced mobility, functionality, and quality of life. Innovations in materials science, robotics, neurological integration, and 3D printing have led to the development of sophisticated prosthetics that closely mimic natural limbs. As technology continues to evolve, future directions will likely focus on advanced sensory feedback, AI integration, biohybrid prosthetics, and increased accessibility. Addressing ethical considerations and social implications will be crucial in ensuring that these advancements benefit all individuals equitably.

The ongoing development of prosthetic limbs holds great promise for further enhancing human capabilities and improving the lives of those with limb loss.

4.1.1 Historical Development of Prosthetic Limbs

Early Prosthetics

The history of prosthetic limbs dates back thousands of years, with the earliest known prosthesis found on a mummified body in ancient Egypt. These early devices were primarily designed for basic functionality, often made from wood and metal, and aimed to restore a semblance of normality to the user's appearance and daily activities.

Post-World War Advancements

Significant advancements in prosthetic technology occurred after World War II, driven by the need to rehabilitate injured soldiers. The introduction of new materials such as plastics and lightweight metals, along with improved manufacturing techniques, led to more functional and comfortable prosthetic limbs. During this period, the focus shifted towards enhancing mobility and usability, with innovations such as adjustable sockets and more natural limb movements.

4.1.2 Technological Advancements in Prosthetic Limbs

Materials Science

Modern prosthetics benefit greatly from advancements in materials science. The use of carbon fiber and other lightweight, durable materials has made prosthetic limbs stronger and more comfortable to wear. These materials also allow for more dynamic movement, mimicking the natural function of biological limbs. Additionally, silicone and other soft, flexible materials are used in sockets to improve the fit and reduce skin irritation.

Robotics and Microprocessors

Robotic technology has revolutionized prosthetic limbs, introducing powered prosthetics that can perform complex movements. Microprocessors and sensors embedded in these devices enable real-time adjustments to the limb's movement, improving balance, coordination, and the overall user experience. For example, advanced microprocessor-controlled knees (MPKs) can adapt to different walking speeds and terrains, providing a more natural gait.

Neurological Integration

One of the most significant advancements in prosthetic limbs is the integration of neurological controls. Myoelectric prosthetics use electrical signals generated by the user's muscles to control the limb's movements. Recent developments in brain-computer interfaces (BCIs) allow for even more direct control, where neural signals from the brain can operate the prosthetic. This level of integration offers a higher degree of functionality and responsiveness, closely mimicking natural limb movements.

3D Printing

The advent of 3D printing technology has made it possible to create customized prosthetic limbs that perfectly fit the user's anatomy. This not only enhances comfort but also reduces production costs and time. 3D printing allows for rapid prototyping and the ability to easily make adjustments based on user feedback. This technology has been particularly beneficial in creating prosthetics for children, who require frequent adjustments as they grow.

4.1.3 Impact of Advancements on Users

Enhanced Mobility and Functionality

The primary benefit of advanced prosthetic limbs is the significant improvement in mobility and functionality. Modern prosthetics enable users to perform a wide range of activities, from walking and running to more intricate tasks like typing or playing a musical instrument. The enhanced functionality provided by microprocessor-controlled joints and myoelectric controls allows for more natural and efficient movement.

Improved Quality of Life

Advancements in prosthetic technology have dramatically improved the quality of life for individuals with limb loss. Enhanced mobility and functionality contribute to greater independence, allowing users to participate more fully in work, social, and recreational activities. The psychological benefits of improved prosthetics, such as increased self-esteem and reduced feelings of disability, are also significant.

Customization and Comfort

The use of advanced materials and 3D printing technology has greatly improved the customization and comfort of prosthetic limbs. Custom-fitted sockets reduce the risk of pressure sores and other complications, while lightweight materials make the prosthetics easier to wear for extended periods. This level of comfort and customization is essential for the long-term success and usability of prosthetic limbs.

Accessibility and Affordability

While advanced prosthetics are often associated with high costs, innovations such as 3D printing have made it possible to produce affordable prosthetic limbs. This has increased accessibility for individuals in developing countries or those with limited financial resources. Organizations and initiatives that focus on low-cost prosthetics are helping to bridge the gap and provide essential mobility solutions to a broader population.

4.1.4 Future Directions in Prosthetic Limb Development

Advanced Sensory Feedback

Future advancements in prosthetic limbs are likely to focus on enhancing sensory feedback. Currently, most prosthetics lack the ability to provide tactile sensations, which limits the user's ability to perceive and interact with their environment. Research is underway to develop sensory feedback systems that can transmit touch, temperature, and pressure sensations to the user's nervous system, offering a more natural and intuitive experience.

Integration with AI

Artificial Intelligence (AI) has the potential to revolutionize prosthetic limbs by providing adaptive and predictive capabilities. AI can analyze the user's movement patterns and make real-time adjustments to the prosthetic's function, improving balance, efficiency, and comfort. Machine learning algorithms can also help in personalizing the prosthetic's responses to better match the user's unique needs and preferences.

Regenerative Medicine and Biohybrid Prosthetics

The future of prosthetics may also involve the integration of regenerative medicine and biohybrid technologies. Researchers are exploring ways to use stem cells and tissue engineering to create biohybrid limbs that can grow and adapt with the user's body. This approach could potentially lead to prosthetics that are seamlessly integrated with the user's biological tissues, offering unprecedented levels of functionality and sensory feedback.

Increased Accessibility and Global Distribution

As technology continues to advance, efforts to make prosthetic limbs more accessible and affordable will be crucial. Initiatives aimed at providing low-cost, high-quality prosthetics to underserved populations will help ensure that more individuals can benefit from these advancements. Collaboration between governments, non-profits, and the private sector will be essential in achieving global distribution and accessibility.

4.1.5 Ethical Considerations and Social Implications

Equity and Accessibility

One of the primary ethical considerations in the development and distribution of advanced prosthetic limbs is ensuring equity and accessibility. While technological advancements have made prosthetics more functional, they have also increased costs, potentially widening the gap between those who can afford advanced prosthetics and those who cannot. Efforts must be made to ensure that these life-changing devices are available to all individuals, regardless of socioeconomic status.

Privacy and Security

As prosthetic limbs become more integrated with digital technologies, concerns about privacy and security arise. Myoelectric and BCI-controlled prosthetics often involve the collection and transmission of sensitive biological data. Ensuring the privacy and security of this data is crucial to protect users from potential misuse or breaches.

Psychological and Social Impacts

The psychological and social impacts of advanced prosthetic limbs must also be considered. While these devices can significantly improve the quality of life for users, they can also create new challenges, such as body image issues or societal perceptions of disability and enhancement. Addressing these psychological and social implications through counseling, education, and societal awareness is essential.

4.2 BIONIC EYES AND HEARING AIDS

Bionic eyes and hearing aids represent groundbreaking advancements in medical technology, offering transformative solutions for individuals with visual and auditory impairments. These innovative devices leverage cutting-edge technologies such as artificial intelligence (AI), robotics, and neuroengineering to restore or enhance sensory capabilities, significantly improving quality of life.

Bionic eyes and hearing aids exemplify the transformative impact of technology on healthcare, offering innovative solutions to restore or enhance sensory capabilities for individuals with visual and auditory impairments. Advances in artificial intelligence, robotics, neuroengineering, and digital signal processing have driven the development of these devices, improving functionality, customization, and user experience.

The integration ofbionic eyes and hearing aids into clinical practice has significantly improved the quality of life for users by restoring sensory function, enhancing communication, and fostering independence. Looking ahead, continued research and development will further enhance these technologies, paving the way for more personalized, effective, and accessible sensory augmentation solutions. Ethical considerations, including affordability, accessibility, and privacy protection, will play a crucial role in shaping the future landscape of bionic eyes, hearing aids, and sensory augmentation technologies as they continue to evolve.

4.2.1 Development of Bionic Eyes

Early Innovations

The development of bionic eyes, or visual prostheses, began with early innovations in the late 20th century. The first successful retinal implants were introduced to restore vision in individuals with severe retinal degenerative diseases such as retinitis pigmentosa and age-related macular degeneration (AMD). These early devices aimed to bypass damaged retinal cells and directly stimulate the remaining functional cells or optic nerves to transmit visual information to the brain.

Retinal Implants and Electrode Arrays

Modern bionic eyes rely on retinal implants equipped with electrode arrays that interface with the retina. These arrays convert light signals captured by an external camera into electrical impulses that stimulate the remaining retinal cells. The Argus II Retinal Prosthesis System, developed by Second Sight Medical Products, was one of the first FDA-approved retinal implants, demonstrating significant advancements in restoring partial vision to blind individuals.

Advances in Neuroengineering

Recent advances in neuroengineering have further refined bionic eyes by improving electrode designs, enhancing biocompatibility, and optimizing stimulation patterns. High-density electrode arrays and advanced signal processing algorithms enable finer spatial resolution and more naturalistic visual perception. Ongoing research aims to integrate wireless communication and AI algorithms to enhance the functionality and adaptability of bionic eyes in various lighting conditions and visual tasks.

4.2.2 Technologies behind Bionic Eyes

Artificial Intelligence and Image Processing

Artificial intelligence plays a crucial role in bionic eyes by processing visual data captured by external cameras. AI algorithms analyze images in real-time, enhancing contrast, recognizing objects, and improving scene interpretation. This enables users to perceive their surroundings more effectively and perform daily tasks such as reading text, recognizing faces, and navigating environments with greater independence.

Robotics and Miniaturization

Bionic eyes benefit from advancements in robotics and miniaturization, allowing for the development of compact, implantable devices that interface seamlessly with the human visual system. Miniature cameras, microprocessors, and wireless communication modules enable discreet integration within the eye, minimizing discomfort and optimizing visual performance.

4.2.3 Applications of Bionic Eyes

Vision Restoration and Enhancement

The primary application of bionic eyes is to restore vision or enhance residual vision in individuals with severe visual impairments. Retinal implants and visual prostheses provide hope to those affected by conditions that cause irreversible damage to the retina, offering the potential to regain functional vision and improve quality of life.

Medical and Surgical Training

Bionic eyes also have applications in medical and surgical training, providing realistic simulations for ophthalmologists and optometrists. These devices allow healthcare professionals to practice complex procedures, such as retinal surgeries and diagnostic assessments, in a controlled and safe environment. Virtual reality

(VR) simulations integrated with bionic eyes enhance training by replicating various retinal pathologies and surgical scenarios.

4.2.4 Impact of Bionic Eyes

Improved Quality of Life

Bionic eyes significantly improve the quality of life for individuals with visual impairments by restoring independence and enhancing daily activities. Users can regain the ability to recognize faces, read text, and navigate unfamiliar environments, reducing dependency on caregivers and improving social interactions.

Psychological and Social Benefits

Beyond functional improvements, bionic eyes offer psychological and social benefits by boosting self-confidence and reducing feelings of isolation. Restored vision allows users to participate more fully in work, education, and recreational activities, fostering a sense of empowerment and inclusion within their communities.

4.2.5 Development of Hearing Aids

Evolution from Analog to Digital

The development of hearing aids has undergone a transformative evolution, transitioning from analog devices that amplify sound to sophisticated digital systems that enhance auditory perception. Early hearing aids used basic amplification technology to amplify incoming sounds, improving hearing but often lacking customization and clarity.

Digital Signal Processing

The advent of digital signal processing (DSP) revolutionized hearing aids by enabling advanced signal analysis and customization. DSP algorithms analyze incoming sound waves in real-time, selectively amplifying frequencies based on the user's hearing profile and environmental conditions. This allows for clearer speech comprehension, noise reduction, and feedback cancellation, enhancing the user's auditory experience.

4.2.6 Technologies behind Hearing Aids

Wireless Connectivity and Smart Features

Modern hearing aids incorporate wireless connectivity and smart features to improve usability and functionality. Bluetooth-enabled hearing aids can wirelessly stream audio from smartphones, televisions, and other devices directly to the user's ears. Smartphone apps allow users to adjust settings, monitor battery life, and customize hearing profiles for different listening environments.

Directional Microphones and Noise Reduction

Directional microphones and noise reduction algorithms are essential technologies in hearing aids, improving speech understanding in noisy environments. Directional microphones focus on capturing sound from specific directions, while noise reduction algorithms suppress background noise, enhancing the clarity of speech and reducing listening effort.

4.2.7 Applications of Hearing Aids

Hearing Rehabilitation

The primary application of hearing aids is in hearing rehabilitation for individuals with mild to severe hearing loss. These devices amplify sound and enhance speech intelligibility, enabling users to communicate effectively and participate in conversations, both in quiet and noisy settings.

Tinnitus Management

Hearing aids equipped with tinnitus masking features provide relief for individuals experiencing tinnitus, a ringing or buzzing sensation in the ears. These devices generate soothing sounds or white noise that mask the perception of tinnitus, helping users manage and alleviate symptoms.

4.2.8 Impact of Hearing Aids

Enhanced Communication and Social Interaction

Hearing aids significantly enhance communication and social interaction by improving speech comprehension and reducing auditory barriers. Users can engage more comfortably in conversations, participate in social activities, and maintain relationships with family and friends, contributing to overall well-being and quality of life.

Cognitive Benefits

There is growing evidence that treating hearing loss with hearing aids can have cognitive benefits. Improved auditory input and communication abilities may help reduce the cognitive load associated with hearing impairment, potentially preserving cognitive function and reducing the risk of cognitive decline over time.

4.2.9 Future Directions in Sensory Augmentation

Integration of AI and Machine Learning

Future advancements in bionic eyes and hearing aids will likely focus on integrating AI and machine learning to enhance performance and customization. AI algorithms can optimize visual processing in bionic eyes, improve speech recognition in noisy environments for hearing aids, and personalize user experiences based on individual preferences and lifestyle.

Biohybrid and Regenerative Approaches

Researchers are exploring biohybrid and regenerative approaches to enhance sensory augmentation devices. Biohybrid bionic eyes may incorporate biological components, such as light-sensitive proteins or neural tissue, to improve visual perception and biocompatibility. Similarly, regenerative strategies aim to restore auditory function by regenerating damaged hair cells or neural pathways in the inner ear.

Ethical Considerations and Accessibility

As bionic eyes and hearing aids become more advanced and accessible, ethical considerations regarding affordability, equitable access, and privacy protection will become increasingly important. Ensuring that these technologies are available to all individuals, regardless of socioeconomic status, and safeguarding user data and privacy will be critical in promoting ethical and responsible adoption.

4.3 THE INTEGRATION OF HUMAN AND MACHINE

The integration of human and machine represents a pivotal frontier in technological advancement, where the boundaries between biology and technology are increasingly blurred. Human-machine integration represents a transformative paradigm shift in technology, bridging the gap between biology and artificial intelligence to enhance human capabilities and quality of life.

From AI-driven automation and robotics to brain-computer interfaces and biohybrid systems, integrated technologies are reshaping industries, healthcare, and societal norms. While the potential benefits of human-machine integration are vast, ethical considerations, including privacy protection, autonomy, and societal equity, must be carefully addressed to ensure responsible innovation and positive societal impact. Embracing collaborative approaches, ethical guidelines, and continuous dialogue will guide the future development and adoption of integrated technologies, paving the way for a more interconnected and technologically advanced society.

4.3.1 Historical Roots of Human-Machine Integration

Early Concepts and Influences

The idea of human-machine integration can be traced back to ancient civilizations, where simple tools extended human capabilities. The Industrial Revolution marked a significant milestone with the mechanization of labor, laying the groundwork for the symbiotic relationship between humans and machines in modern society. The concept gained prominence in the 20th century with advancements in cybernetics, cognitive science, and artificial intelligence (AI), shaping the vision of intelligent machines collaborating with humans.

4.3.2 Technologies Facilitating Integration

Artificial Intelligence (AI) and Machine Learning

Artificial intelligence and machine learning play a pivotal role in human-machine integration by enabling machines to learn from data, make decisions, and perform tasks autonomously. AI algorithms are increasingly integrated into various applications, from personal assistants and autonomous vehicles to medical diagnostics and industrial automation. Machine learning models continuously improve their performance through iterative learning processes, enhancing their ability to interact and collaborate with humans effectively.

Robotics and Automation

Robotics and automation technologies enhance human-machine integration by automating repetitive tasks and augmenting human capabilities in various industries. Collaborative robots (cobots) work alongside human workers, assisting with assembly, logistics, and hazardous operations. Advances in exoskeletons and prosthetics further blur the line between human and machine, enhancing strength, mobility, and endurance in industrial, medical, and military contexts.

Brain-Computer Interfaces (BCIs)

Brain-computer interfaces (BCIs) enable direct communication between the human brain and external devices, facilitating seamless interaction and control. BCIs translate neural signals into commands that operate computers, prosthetics, or assistive devices, offering new possibilities for individuals with disabilities and enhancing human performance in fields such as gaming and neurorehabilitation. Ongoing research aims to improve BCI accuracy, bandwidth, and biocompatibility for broader applications.

4.3.3 Ethical Considerations in Human-Machine Integration

Privacy and Data Security

The integration of human and machine raises concerns about privacy and data security, particularly with the collection and utilization of personal data by AI systems and BCIs. Protecting sensitive information from unauthorized access, ensuring transparency in data usage, and establishing robust cybersecurity measures are essential to safeguarding user privacy and maintaining trust in integrated technologies.

Autonomy and Control

As machines become more autonomous and capable of independent decision-making, ethical considerations arise regarding human oversight, accountability, and control. Ensuring that humans retain the ability to intervene, override, or deactivate autonomous systems is crucial for mitigating risks and maintaining ethical standards in human-machine interactions.

Equity and Accessibility

Ensuring equitable access to integrated technologies is essential to prevent exacerbating societal inequalities. Addressing disparities in access to AI-driven healthcare, educational resources, and employment opportunities requires proactive measures to promote inclusivity and reduce digital divides.

Making integrated technologies accessible to marginalized communities and developing regions can foster socioeconomic empowerment and enhance global equity.

4.3.4 Societal Implications of Human-Machine Integration

Workforce Transformation

Human-machine integration is reshaping the global workforce by augmenting human capabilities and automating routine tasks. While automation enhances productivity and efficiency in industries such as manufacturing and logistics, it also necessitates workforce reskilling and upskilling to adapt to changing job roles and technological advancements.

Collaborative work environments, where humans and machines complement each other's strengths, are becoming increasingly prevalent.

Healthcare and Well-being

Integrated technologies hold transformative potential in healthcare, enabling personalized medicine, remote diagnostics, and therapeutic interventions. AI-driven healthcare systems can analyze vast amounts of patient data to improve diagnostics, treatment planning, and drug discovery.

Assistive technologies, such as robotic prosthetics and smart implants, enhance mobility and quality of life for individuals with disabilities, highlighting the positive impact of human-machine integration on healthcare outcomes.

Ethical and Social Norms

Human-machine integration challenges traditional ethical and social norms regarding identity, autonomy, and the definition of being human. Debates surrounding AI ethics, robot rights, and the implications of BCIs on cognitive enhancement raise philosophical questions about the ethical boundaries of technological intervention in human life.

Balancing technological advancement with ethical considerations requires continuous dialogue, regulatory frameworks, and ethical guidelines to ensure responsible innovation and societal acceptance.

4.3.5 Future Directions in Human-Machine Integration

Cognitive Augmentation and AI Collaboration

Future advancements in human-machine integration will focus on enhancing cognitive capabilities through AI collaboration and neural augmentation. Integrating AI assistants into everyday tasks, such as decision-making, problem-solving, and information retrieval, can amplify human cognitive abilities and productivity. Research in neural interfaces aims to develop seamless brain-to-machine communication for enhancing memory, learning, and sensory perception.

Biohybrid Systems and Regenerative Medicine

Biohybrid systems and regenerative medicine offer promising avenues for integrating biological and synthetic components to restore or enhance human functions. Bioengineered tissues, organs, and neural interfaces could revolutionize healthcare by providing biocompatible solutions for organ replacement, sensory augmentation, and neural repair. Advances in tissue engineering and biocompatible materials will facilitate the development of biohybrid systems with improved functionality and longevity.

Ethical Guidelines and Governance

As human-machine integration technologies advance, establishing robust ethical guidelines, governance frameworks, and regulatory standards will be crucial to ensure responsible development, deployment, and use. Collaboration between policymakers, industry stakeholders, ethicists, and the public is essential for addressing ethical dilemmas, safeguarding human rights, and promoting equitable access to integrated technologies. Proactive measures to anticipate and mitigate potential risks, such as algorithmic bias, data privacy breaches, and unintended consequences, are essential for fostering trust and ethical adoption of human-machine integration technologies.

CHAPTER 5

NEUROENHANCEMENT AND COGNITIVE COMPUTING

5.1 ENHANCING MEMORY AND COGNITIVE SKILLS

Enhancing memory and cognitive skills has been a longstanding goal of humanity, driven by our innate desire to optimize mental abilities and achieve peak cognitive performance. Enhancing memory and cognitive skills is a multifaceted endeavor encompassing lifestyle factors, cognitive training techniques, technological innovations, and ethical considerations. From physical exercise and mindfulness practices to brain training apps and neurostimulation technologies, diverse approaches offer opportunities to optimize cognitive function and support cognitive health across the lifespan. Ethical guidelines and regulatory frameworks play a pivotal role in guiding the development, deployment, and ethical use of cognitive enhancement technologies, balancing potential benefits with ethical concerns and societal implications. As research advances and technologies evolve, fostering interdisciplinary collaboration and public engagement will shape the future landscape of cognitive enhancement, promoting responsible innovation and equitable access to enhance cognitive abilities and improve quality of life.

5.1.1 Understanding Memory and Cognitive Skills

Memory Processes

Memory encompasses the ability to encode, store, and retrieve information. It involves multiple processes, including sensory memory, short-term memory, and long-term memory. Encoding transforms sensory information into a form that can be stored, while retrieval retrieves stored information for conscious awareness and use. Cognitive skills, on the other hand, refer to mental abilities such as attention, problem-solving, reasoning, and decision-making that enable effective processing and utilization of information.

5.1.2 Lifestyle Factors for Cognitive Enhancement

Physical Exercise

Physical exercise has been shown to benefit cognitive function by promoting neuroplasticity, enhancing blood flow to the brain, and stimulating the release of neurotransmitters such as dopamine and serotonin. Regular aerobic exercise improves memory, attention, and executive functions, potentially reducing the risk of cognitive decline and neurodegenerative diseases.

Diet and Nutrition

A balanced diet rich in antioxidants, omega-3 fatty acids, and vitamins (e.g., B vitamins, vitamin D) supports brain health and cognitive function. Foods such as fatty fish, fruits, vegetables, nuts, and whole grains provide essential nutrients that protect against oxidative stress, inflammation, and age-related cognitive decline. Maintaining hydration and avoiding excessive alcohol consumption also contribute to optimal brain function.

Sleep Quality

Quality sleep is crucial for memory consolidation and cognitive performance. During sleep, the brain processes and stores newly acquired information, strengthening synaptic connections and enhancing learning. Chronic sleep deprivation impairs memory, attention, and decision-making abilities, highlighting the importance of prioritizing adequate sleep for cognitive enhancement.

5.1.3 Cognitive Training Techniques

Brain Training Apps and Games

Brain training apps and games designed to challenge memory, attention, and problem-solving skills have gained popularity as cognitive enhancement tools. These digital platforms offer interactive exercises and puzzles that stimulate neural pathways, improve cognitive flexibility, and enhance mental agility. While some studies suggest potential benefits from brain training, the efficacy and transferability of gains to real-world tasks remain debated among researchers.

Mnemonic Strategies

Mnemonic techniques involve using memory aids, such as acronyms, visual imagery, and storytelling, to enhance information retention and recall. These strategies leverage the brain's ability to associate new information with pre-existing knowledge or vivid mental images, facilitating easier encoding and retrieval processes. Mnemonic devices are widely used in education, language learning, and professional settings to improve memorization and learning efficiency.

Meditation

Meditation is a practice in which an individual uses a technique – such as mindfulness, or focusing the mind on a particular object, thought, or activity – to train attention and awareness, and achieve a mentally clear and emotionally calm and stable state.

Meditation is practiced in numerous religious traditions. The earliest records of meditation (*dhyana*) are found in the Upanishads, and meditation plays a salient role in the contemplative repertoire of Jainism, Buddhism and Hinduism. Since the 19th century, Asian meditative techniques have spread to other cultures where they have also found application in non-spiritual contexts, such as business and health.

Meditation may significantly reduce stress, anxiety, depression, and pain, and enhance peace, perception, self-concept, and well-being. Research is ongoing to better understand the effects of meditation on health (psychological, neurological, and cardiovascular) and other areas.

No Universally Accepted Definition

Meditation has proven difficult to define as it covers a wide range of dissimilar practices in different traditions. In popular usage, the word "meditation" and the phrase "meditative practice" are often used imprecisely to designate practices found across many cultures. These can include almost anything that is claimed to train the attention of mind or to teach calmness or compassion. There remains no definition of necessary and sufficient criteria for meditation that has achieved widespread acceptance within the modern scientific community.

Separation of Technique from Tradition

Some of the difficulty in precisely defining meditation has been in recognizing the particularities of the many various traditions; and theories and practice can differ within a tradition. Taylor noted that even within a faith such as "Hindu" or "Buddhist", schools and individual teachers may teach distinct types of meditation. Ornstein noted that "Most techniques of meditation do not exist as solitary practices but are only artificially separable from an entire system of practice and belief." For instance, while monks meditate as part of their everyday lives, they also engage in the codified rules and live together in monasteries in specific cultural settings that go along with their meditative practices.

Dictionary Definitions

Dictionaries give both the original Latin meaning of "think[ing] deeply about (something)"; as well as the popular usage of "focusing one's mind for a period of time", "the act of giving your attention to only one thing, either as a religious activity or as a way of becoming calm and relaxed", and "to engage in mental exercise (such as concentrating on one's breathing or repetition of a mantra) for the purpose of reaching a heightened level of spiritual awareness."

Scholarly Definitions

In modern psychological research, meditation has been defined and characterized in various ways. Many of these emphasize the role of attention and characterize the practice of meditation as attempts to get beyond the reflexive, "discursive thinking" or "logic" mind to achieve a deeper, more devout, or more relaxed state.

Bond et al. (2009) identified criteria for defining a practice as meditation "for use in a comprehensive systematic review of the therapeutic use of meditation", using "a 5-round Delphi study with a panel of 7 experts in meditation research" who were also trained in diverse but empirically highly studied (Eastern-derived or clinical) forms of meditation:

- three main criteria ... as essential to any meditation practice: the use of a defined technique, logic relaxation, and a self-induced state/mode.
- Other criteria deemed important [but not essential] involve a state of psychophysical relaxation, the use of a self-focus skill or anchor, the presence of a state of suspension of logical thought processes, a religious/spiritual/philosophical context, or a state of mental silence.
- ... It is plausible that meditation is best thought of as a natural category of techniques best captured by 'family resemblances' ... or by the related 'prototype' model of concepts."

Several other definitions of meditation have been used by influential modern reviews of research on meditation across multiple traditions:

- **Walsh & Shapiro (2006):** "Meditation refers to a family of self-regulation practices that focus on training attention and awareness in order to bring mental processes under greater voluntary control and thereby foster general mental well-being and development and/or specific capacities such as calm, clarity, and concentration"
- **Cahn & Polich (2006):** "*Meditation* is used to describe practices that self-regulate the body and mind, thereby affecting mental events by engaging a specific attentional set.... regulation of attention is the central commonality across the many divergent methods"
- **Jevning et al. (1992):** "We define meditation... as a stylized mental technique... repetitively practiced for the purpose of attaining a subjective experience that is frequently described as very restful, silent, and of heightened alertness, often characterized as blissful"
- **Goleman (1988):** "The need for the meditator to retrain his attention, whether through concentration or mindfulness, is the single invariant ingredient in... every meditation system"

Classifications

In the West, meditation techniques have often been classified in two broad categories, which in actual practice are often combined: focused (or concentrative) meditation and open monitoring (or mindfulness) meditation:

- Direction of mental attention... A practitioner can focus intensively on one particular object (so-called *concentrative meditation*), on all mental events that enter the field of awareness (so-called *mindfulness meditation*), or both specific focal points and the field of awareness.

Focused methods include paying attention to the breath, to an idea or feeling (such as *mettā* – loving-kindness), to a *kōan*, or to a *mantra* (such as in transcendental meditation), and single point meditation. Open monitoring methods include mindfulness, *shikantaza* and other awareness states.

Another typology divides meditation approaches into concentrative, generative, receptive and reflective practices:

- concentrative: focused attention, including breath meditation, TM, and visualizations;
- generative: developing qualities like loving kindness and compassion;
- receptive: open monitoring; and
- reflective: systematic investigation, contemplation.

The Buddhist tradition often divides meditative practice into *samatha*, or calm abiding, and *vipassana*, insight. Mindfulness of breathing, a form of focused attention, calms down the mind; this calmed mind can then investigate the nature of reality, by monitoring the fleeting and ever-changing constituents of experience, by reflective investigation, or by "turning back the radiance," focusing awareness on awareness itself and discerning the true nature of mind as awareness itself.

Matko and Sedlmeier (2019) "call into question the common division into "focused attention" and "open-monitoring" practices." They argue for "two orthogonal dimensions along which meditation techniques could be classified," namely "activation" and "amount of body orientation," proposing seven clusters of techniques: "mindful observation, body-centered meditation, visual concentration, contemplation, affect-centered meditation, mantra meditation, and meditation with movement." Jonathan Shear argues that transcendental meditation is an "automatic self-transcending" technique, different from focused attention and open monitoring. In this kind of practice, "there is no attempt to sustain any particular condition at all. Practices of this kind, once started, are reported to automatically "transcend" their own activity and disappear, to be started up again later if appropriate." Yet, Shear also states that "automatic self-transcending" also applies to the way other techniques such as from Zen and Qigong are practiced by experienced meditators "once they had become effortless and automatic through years of practice."

Technique

Asanas or body postures such as *padmasana* (full-lotus, half-lotus), cross-legged sitting, *seiza*, and kneeling positions are popular meditative postures in Hinduism, Buddhism and Jainism, although other postures such as sitting, supine (lying), and standing are also used. Meditation is also sometimes done while walking, known as *kinhin*, while doing a simple task mindfully, known as *samu*, or while lying down, known as *shavasana*.

The Transcendental Meditation technique recommends practice of 20 minutes twice per day. Some techniques suggest less time, especially when starting meditation, and Richard Davidson has quoted research saying benefits can be achieved with a practice of only 8 minutes per day. Research shows improvement in meditation time with simple oral and video training. Some meditators practice for much longer, particularly when on a course or retreat. Some meditators find practice best in the hours before dawn.

Some religions have traditions of using prayer beads as tools in devotional meditation. Most prayer beads and Christian rosaries consist of pearls or beads linked together by a thread. The Roman Catholic rosary is a string of beads containing five sets with ten small beads. Eastern and Oriental Orthodox have traditions of using prayer ropes called Comboschini or Meqetaria as an aid to prayerful meditation.

The Hindu *japa mala* has 108 beads (the figure 108 in itself having spiritual significance as the energy of the sounds equivalates to Om), as well as those used in Gaudiya Vaishnavism, the Hare Krishna tradition, Jainism and Buddhist prayer beads. Each bead is counted once as a person recites a mantra until the person has gone all the way around the mala.

The Muslim *misbaha* has 99 beads. There is also quite a variance when it comes to materials used for beads. Beads made from seeds of rudraksha trees are considered sacred by devotees of Shiva, while followers of Vishnu revere the wood that comes from the tulsi plant.

The Buddhist literature has many stories of Enlightenment being attained through disciples being struck by their masters. According to T. Griffith Foulk, the encouragement stick was an integral part of the Zen practice:

- In the Rinzai monastery where I trained in the mid-1970s, according to an unspoken etiquette, monks who were sitting earnestly and well were shown respect by being hit vigorously and often; those known as laggards were ignored by the hall monitor or given little taps if they requested to be hit. Nobody asked about the 'meaning' of the stick, nobody explained, and nobody ever complained about its use.

Neuroscientist and long-time meditator Richard Davidson has expressed the view that having a narrative can help the maintenance of daily practice. For instance he himself prostrates to the teachings, and meditates "not primarily for my benefit, but for the benefit of others".

Meditation Traditions

The history of meditation is intimately bound up with the religious context within which it was practiced. Rossano has suggested that the emergence of the capacity for focused attention, an element of many methods of meditation, may have contributed to the latest phases of human biological evolution. Some of the earliest references to meditation, as well as proto-Samkhya, are found in the Upanishads of India. According to Wynne, the earliest clear references to meditation are in the middle Upanishads and the Mahabharata (including the Bhagavad Gita). According to Gavin Flood, the earlier Brihadaranyaka Upanishad is describing meditation when it states that "having become calm and concentrated, one perceives the self (ātman) within oneself" (BU 4.4.23).

Jain meditation and spiritual practices system were referred to as salvation-path. It has three parts called the Ratnatraya "Three Jewels": right perception and faith, right knowledge and right conduct. Meditation in Jainism aims at realizing the self, attaining salvation, and taking the soul to complete freedom. It aims to reach and to remain in the pure state of soul which is believed to be pure consciousness, beyond any attachment or aversion. The practitioner strives to be just a knower-seer (gyata-drashta). Jain meditation can be broadly categorized to Dharma Dhyana and Shukla Dhyana.

Jainism uses meditation techniques such as pindāstha-dhyāna, padāstha-dhyāna, rūpāstha-dhyāna, rūpātita-dhyāna, and savīrya-dhyāna. In padāstha dhyāna one focuses on a mantra. A mantra could be either a combination of core letters or words on deity or themes. There is a rich tradition of mantra in Jainism. All Jain followers irrespective of their sect, whether Digambara or Śvetāmbara, practice mantra. Mantra chanting is an important part of daily lives of Jain monks and followers. Mantra chanting can be done either loudly or silently in mind.

Contemplation is a very old and important meditation technique. The practitioner meditates deeply on subtle facts. In agnya vichāya, one contemplates on seven facts – life and non-life, the inflow, bondage, stoppage and removal of karmas, and the final accomplishment of liberation. In apaya vichāya, one contemplates on the incorrect insights one indulges, which eventually develops right insight. In vipaka vichāya, one reflects on the eight causes or basic types of karma. In sansathan vichāya, one thinks about the vastness of the universe and the loneliness of the soul.

Buddhists pursue meditation as part of the path toward awakening and nirvana. The closest words for meditation in the classical languages of Buddhism are bhāvanā ("development"), and the core practices of body contemplations (repulsiveness and cemetery contemplations) and anapanasati (mindfulness of in-and-out breathing) culminating in jhāna/dhyāna or samādhi.

While most classical and contemporary Buddhist meditation guides are school-specific, the root meditative practices of various body recollections and breath meditation have been preserved and transmitted in almost all Buddhist traditions, through Buddhist texts like the Satipatthana Sutta and the Dhyana sutras, and through oral teacher-student transmissions. These ancient practices are supplemented with various distinct interpretations of, and developments in, these practices.

The Theravāda tradition stresses the development of samatha and vipassana, postulating over fifty methods for developing mindfulness based on the Satipatthana Sutta, and forty for developing concentration based on the Visuddhimagga.

The Tibetan tradition incorporated Sarvastivada and Tantric practices, wedded with Madhyamaka philosophy, and developed thousands of visualization meditations.

The Silk Road transmission of Buddhism introduced Buddhist meditation to other Asian countries, reaching China in the 2nd century CE, and Japan in the 6th century CE. In the modern era, Buddhist meditation techniques have become popular in the wider world, due to the influence of Buddhist modernism on Asian Buddhism, and western lay interest in Zen and the Vipassana movement, with many non-Buddhists taking-up meditative practices. The modernized concept of mindfulness (based on the Buddhist term sati) and related meditative practices have in turn led to mindfulness based therapies.

Dhyana may have been an original contribution of Gautama Buddha (5th cent. BCE), the founder of Buddhism. While often presented as a form of focused attention or concentration, as in Buddhagosa's Theravada classic the Visuddhimagga ("Path of purification, 5th c. CE), according to a number contemporary scholars and scholar-practitioners, it is actually a description of the development of perfected equanimity and mindfulness, apparently induced by satipatthana, an open monitoring of the breath, without trying to regulate it. The same description, in a different formula, can be found in the bojjhanga, the "seven factors of awakening," and may therefore refer to the core program of early Buddhist bhavana. According to Vetter, dhyana seems to be a natural development from the sense-restraint and moral constrictions prescribed by the Buddhist tradition.

The Buddha identified two paramount mental qualities that arise from wholesome meditative practice or bhavana, namely samatha ("calm," "serenity" "tranquility") and vipassana (insight). As the developing tradition started to emphasize the value of liberating insight, and dhyana came to be understood as concentration, samatha and vipassana were understood as two distinct meditative techniques. In this understanding, samatha steadies, composes, unifies and concentrates the mind, while vipassana enables one to see, explore and discern "formations" (conditioned phenomena based on the five aggregates).

According to this understanding, which is central to Theravada orthodoxy but also plays a role in Tibetan Buddhism, through the meditative development of serenity, one is able to weaken the obscuring hindrances and bring the mind to a collected, pliant, and still state (samadhi). This quality of mind then supports the development of insight and wisdom (Prajñā) which is the quality of mind that can "clearly see" (vi-passana) the nature of phenomena. What exactly is to be seen varies within the Buddhist traditions. In Theravada, all phenomena are to be seen as impermanent, suffering, not-self and empty. When this happens, one develops dispassion (viraga) for all phenomena, including all negative qualities and hindrances and lets them go. It is through the release of the hindrances and ending of craving through the meditative development of insight that one gains liberation.

Christian meditation is a term for a form of prayer in which a structured attempt is made to get in touch with and deliberately reflect upon the revelations of God. In the Roman Empire, by 20 BCE Philo of Alexandria had written on some form of "spiritual exercises" involving attention (prosoche) and concentration and by the 3rd century Plotinus had developed meditative techniques. The word meditation comes from the Latin word meditatum, which means to "concentrate" or "to ponder". Monk Guigo II introduced this terminology for the first time in the 12th century AD. Christian meditation is the process of deliberately focusing on specific thoughts (e.g. a biblical scene involving Jesus and the Virgin Mary) and reflecting on their meaning in the context of the love of God. Christian meditation is sometimes taken to mean the middle level in a broad three-stage characterization of prayer: it then involves more reflection than first level vocal prayer, but is more structured than the multiple layers of contemplation in Christianity.

Between the 10th and 14th centuries, hesychasm was developed, particularly on Mount Athos in Greece, and involves the repetition of the Jesus prayer. Interactions with Indians or the Sufis may have influenced the Eastern Christian meditation approach to hesychasm, but this is unproven.

Western Christian meditation contrasts with most other approaches in that it does not involve the repetition of any phrase or action and requires no specific posture. Western Christian meditation progressed from the 6th century practice of Bible reading among Benedictine monks called Lectio Divina, i.e. divine reading. Its four formal steps as a "ladder" were defined by the monk Guigo II in the 12th century with the Latin terms lectio, meditatio, oratio, and contemplatio (i.e. read, ponder, pray, contemplate). Western Christian meditation was further developed by saints such as Ignatius of Loyola and Teresa of Avila in the 16th century.

On April 28, 2021, Pope Francis, in an address to the General Audience, said that meditation is a need for everyone. He noted that the term "meditation" has had many meanings throughout history, and that "the ancients used to say that the organ of prayer is the heart."

In Catholic Christianity, the Rosary is a devotion for the meditation of the mysteries of Jesus and Mary. "The gentle repetition of its prayers makes it an excellent means to moving into deeper meditation. It gives us an opportunity to open ourselves to God's word, to refine our interior gaze by turning our minds to the life of Christ. The first principle is that meditation is learned through practice. Many people who practice rosary meditation begin very simply and gradually develop a more sophisticated meditation. The meditator learns to hear an interior voice, the voice of God. Similarly, the chotki of the Eastern Orthodox denomination, the Wreath of Christ of the Lutheran faith, and the Anglican prayer beads of the Episcopalian tradition are used for Christian prayer and meditation.

According to Edmund P. Clowney, Christian meditation contrasts with Eastern forms of meditation as radically as the portrayal of God the Father in the Bible contrasts with depictions of Krishna or Brahman in Indian teachings. Unlike some Eastern styles, most styles of Christian meditation do not rely on the repeated use of mantras, and yet are also intended to stimulate thought and deepen meaning. Christian meditation aims to heighten the personal relationship based on the love of God that marks Christian communion. In Aspects of Christian meditation, the Catholic Church warned of potential incompatibilities in mixing Christian and Eastern styles of meditation. In 2003, in A Christian reflection on the New Age the Vatican announced that the "Church avoids any concept that is close to those of the New Age".

Dhikr (zikr) is a type of meditation within Islam, meaning remembering and mentioning God, which involves the repetition of the 99 Names of God since the 8th or 9th century. It is interpreted in different meditative techniques in Sufism or Islamic mysticism. This became one of the essential elements of Sufism as it was systematized traditionally. It is juxtaposed with fikr (thinking) which leads to knowledge. By the 12th century, the practice of Sufism included specific meditative techniques, and its followers practiced breathing controls and the repetition of holy words. Sufism uses a meditative procedure like Buddhist concentration, involving high-intensity and sharply focused introspection. In the Oveyssi-Shahmaghsoudi Sufi order, for example, *muraqabah* takes the form of *tamarkoz*, "concentration" in Persian.

Tafakkur or *tadabbur* in Sufism literally means *reflection upon the universe*: this is considered to permit access to a form of cognitive and emotional development that can emanate only from the higher level, i.e. from God. The sensation of receiving divine inspiration awakens and liberates both heart and intellect, permitting such inner growth that the apparently mundane actually takes on the quality of the infinite. Muslim teachings embrace life as a test of one's submission to God. Dervishes of certain Sufi orders practice whirling, a form of physically active meditation.

Meditation has spread in the West since the late 19th century, accompanying increased travel and communication among cultures worldwide. Most prominent has been the transmission of Asian-derived practices to the West. In addition, interest in some Western-based meditative practices has been revived, and these have been disseminated to a limited extent to Asian countries.

Ideas about Eastern meditation had begun "seeping into American popular culture even before the American Revolution through the various sects of European occult Christianity", and such ideas "came pouring in [to America] during the era of the transcendentalists, especially between the 1840s and the 1880s." The following decades saw further spread of these ideas to America:

- The World Parliament of Religions, held in Chicago in 1893, was the landmark event that increased Western awareness of meditation. This was the first time that Western audiences on American soil received Asian spiritual teachings from Asians themselves. Thereafter, Swami Vivekananda [...] [founded] various Vedanta ashrams [...] Anagarika Dharmapala lectured at Harvard on Theravada Buddhist meditation in 1904; Abdul Baha [...] [toured] the US teaching the principles of Bahai [*sic*], and Soyen Shaku toured in 1907 teaching Zen.

More recently, in the 1960s, another surge in Western interest in meditative practices began. The rise of communist political power in Asia led to many Asian spiritual teachers taking refuge in Western countries, oftentimes as refugees. In addition to spiritual forms of meditation, secular forms of meditation have taken root. Rather than focusing on spiritual growth, secular meditation emphasizes stress reduction, relaxation and self-improvement.

The 2012 US National Health Interview Survey (NHIS) (34,525 subjects) found 8% of US adults used meditation, with lifetime and 12-month prevalence of meditation use of 5.2% and 4.1% respectively. In the 2017 NHIS survey, meditation use among workers was 10% (up from 8% in 2002).

Mantra meditation, with the use of a *japa mala* and especially with focus on the Hare Krishna maha-mantra, is a central practice of the Gaudiya Vaishnava faith tradition and the International Society for Krishna Consciousness (ISKCON), also known as the Hare Krishna movement. Other popular New Religious Movements include the Ramakrishna Mission, Vedanta Society, Divine Light Mission, Chinmaya Mission, Osho, Sahaja Yoga, Transcendental Meditation, Oneness University, Brahma Kumaris, Vihangam Yoga and Heartfulness Meditation (Sahaj Marg).

New Age meditations are often influenced by Eastern philosophy, mysticism, yoga, Hinduism and Buddhism, yet may contain some degree of Western influence. In the West, meditation found its mainstream roots through the social revolution of the 1960s and 1970s, when many of the youth of the day rebelled against traditional religion as a reaction against what some perceived as the failure of Christianity to provide spiritual and ethical guidance. New Age meditation as practised by the early hippies is regarded for its techniques of blanking out the mind and releasing oneself from conscious thinking. This is often aided by repetitive chanting of a mantra, or focusing on an object. New Age meditation evolved into a range of purposes and practices, from serenity and balance to access to other realms of consciousness to the concentration of energy in group meditation to the supreme goal of *samadhi*, as in the ancient yogic practice of meditation.

Guided meditation is a form of meditation which uses a number of different techniques to achieve or enhance the meditative state. It may simply be meditation done under the guidance of a trained practitioner or teacher, or it may be through the use of imagery, music, and other techniques.

The session can be either in person, via media comprising music or verbal instruction, or a combination of both. The most common form is a combination of meditation music and receptive music therapy, guided imagery, relaxation, mindfulness, and journaling.

Because of the different combinations used under the one term, it can be difficult to attribute positive or negative outcomes to any of the various techniques. Furthermore, the term is frequently used interchangeably with "guided imagery" and sometimes with "creative visualization" in popular psychology and self-help literature. It is less commonly used in scholarly and scientific publications. Consequently, guided meditation cannot be understood as a single technique but rather multiple techniques that are integral to its practice.

Guided meditation as an aggregate or synthesis of techniques includes meditation music, receptive music therapy, guided imagery, relaxation, meditative praxis, and self-reflective journaling, all of which have been shown to have therapeutic benefits when employed as an adjunct to primary strategies. Benefits include lower levels of stress, reducing asthmatic episodes, physical pain, insomnia, episodic anger, negative or irrational thinking, and anxiety, as well as improving coping skills, focus, and a general feeling of well-being.

Psychotherapy

Carl Jung (1875–1961) was an early western explorer of eastern religious practices. He clearly advocated ways to increase the conscious awareness of an individual. Yet he expressed some caution concerning a westerner's direct immersion in eastern practices without some prior appreciation of the differing spiritual and cultural contexts. Also Erich Fromm (1900–1980) later explored spiritual practices of the east.

Clinical

The US National Center for Complementary and Integrative Health states that *"Meditation and mindfulness practices may have a variety of health benefits and may help people improve the quality of their lives. Recent studies have investigated if meditation or mindfulness helps people manage anxiety, stress, depression, pain, or symptoms related to withdrawal from nicotine, alcohol, or opioids."* However, the NCCIC goes on to caution that, *"results from the studies have been difficult to analyze and may have been interpreted too optimistically."*

A 2014 review found that practice of mindfulness meditation for two to six months by people undergoing long-term psychiatric or medical therapy could produce moderate improvements in anxiety, pain, or depression. In 2017, the American Heart Association issued a scientific statement that meditation may be a reasonable adjunct practice to help reduce the risk of cardiovascular diseases, with the qualification that meditation needs to be better defined in higher-quality clinical research of these disorders. Recent findings have also found evidence of meditation affecting migraines in adults. Mindfulness meditation may allow for a decrease in migraine episodes, and a drop in migraine medication usage.

Low-quality evidence indicates that meditation may help with irritable bowel syndrome, insomnia, cognitive decline in the elderly, and post-traumatic stress disorder. Sitting in silence, body scan meditation and concentrating on breathing was shown in a 2016 review to moderately decrease symptoms of PTSD and depression in war veterans. Researchers have found that participating in mindfulness meditation can aid insomnia patients by improving sleep quality and total wake time. Mindfulness meditation is not a treatment for insomnia patients, but it can provide support in addition to their treatment options.

In the Workplace

A 2010 review of the literature on spirituality and performance in organizations found an increase in corporate meditation programs. As of 2016 around a quarter of U.S. employers were using stress reduction initiatives. The goal was to help reduce stress and improve reactions to stress. Aetna now offers its program to its customers. Google also implements mindfulness, offering more than a dozen meditation courses, with the most prominent one, "Search Inside Yourself", having been implemented since 2007. General Mills offers the Mindful Leadership Program Series, a course which uses a combination of mindfulness meditation, yoga and dialogue with the intention of developing the mind's capacity to pay attention. Many military organizations around the world have found meditation and mindfulness practice can support a range of benefits related to combat, including support for mental health, mental clarity, focus and stress control.

In School

A review of 15 peer-reviewed studies of youth meditation in schools indicated transcendental meditation a moderate effect on wellbeing and a small effect on social competence. Insufficient research has been done on the effect of meditation on academic achievement. Evidence has also shown possible improvement to stress, cognitive performance in school taught meditation. Positive effects on emotion regulation, stress and anxiety can also be seen in students in university and nursing.

Relaxation Response and Biofeedback

Herbert Benson of Harvard Medical School conducted a series of clinical tests on meditators from various disciplines, including the Transcendental Meditation technique and Tibetan Buddhism. In 1975, Benson published a book titled *The Relaxation Response* where he outlined his own version of meditation for relaxation. Also in the 1970s, the American psychologist Patricia Carrington developed a similar technique called Clinically Standardized Meditation (CSM). In Norway, another sound-based method called Acem Meditation developed a psychology of meditation and has been the subject of several scientific studies. Biofeedback has been used by many researchers since the 1950s in an effort to enter deeper states of mind.

Effects

Research on the processes and effects of meditation is a subfield of neurological research. Modern scientific techniques, such as fMRI and EEG, were used to observe neurological responses during meditation. Concerns have been raised on the quality of meditation research, including the particular characteristics of individuals who tend to participate.

Meditation lowers heart rate, oxygen consumption, breathing frequency, stress hormones, lactate levels, and sympathetic nervous system activity (associated with the fight-or-flight response), along with a modest decline in blood pressure. However, those who have meditated for two or three years were found to already have low blood pressure. During meditation, the oxygen consumption decrease averages 10 to 20 percent over the first three minutes. During sleep for example, oxygen consumption decreases around 8 percent over four or five hours. For meditators who have practiced for years, breath rate can drop to three or four breaths per minute and "brain waves slow from the usual beta (seen in waking activity) or alpha (seen in normal relaxation) to much slower delta and theta waves".

Since the 1970s, clinical psychology and psychiatry have developed meditation techniques for numerous psychological conditions. Mindfulness practice is employed in psychology to alleviate mental and physical conditions, such as reducing depression, stress, and anxiety. Mindfulness is also used in the treatment of drug addiction, although the quality of research has been poor. Studies demonstrate that meditation has a moderate effect to reduce pain. There is insufficient evidence for any effect of meditation on positive mood, attention, eating habits, sleep, or body weight.

A 2017 systematic review and meta-analysis of the effects of meditation on empathy, compassion, and prosocial behaviors found that meditation practices had small to medium effects on self-reported and observable outcomes, concluding that such practices can "improve positive prosocial emotions and behaviors". However, a meta-review published on *Scientific Reports* showed that the evidence is very weak and "that the effects of meditation on compassion were only significant when compared to passive control groups suggests that other forms of active interventions (like watching a nature video) might produce similar outcomes to meditation". Another study reported increased ability to correctly report others facial expressions over time after meditating using fMRI. Loving kindness meditation has also been shown to have positive effects on feelings of connection and decreased attention to oneself.

Mindfulness

Mindfulness is the cognitive skill, usually developed through meditation, of sustaining meta-attention of the contents of one's own mind in the present moment. Mindfulness derives from *sati*, a significant element of Hindu and Buddhist traditions, and is based on Zen, *Vipassanā*, and Tibetan meditation techniques. Though definitions and techniques of mindfulness are wide-ranging, Buddhist traditions describe what constitutes mindfulness, such as how perceptions of the past, present and future arise and cease as momentary sense-impressions and mental phenomena. Individuals who have contributed to the popularity of mindfulness in the modern Western context include Thích Nhất Hạnh, Joseph Goldstein, Herbert Benson, Jon Kabat-Zinn, and Richard J. Davidson.

Clinical psychology and psychiatry since the 1970s have developed a number of therapeutic applications based on mindfulness for helping people experiencing a variety of psychological conditions. Mindfulness practice has been employed to reduce depression, stress, anxiety, and in the treatment of drug addiction. Programs based on mindfulness models have been adopted within schools, prisons, hospitals, veterans' centers, and other environments, and mindfulness programs have been applied for additional outcomes such as for healthy aging, weight management, athletic performance, helping children with special needs, and as an intervention during early pregnancy.

Clinical studies have documented both physical- and mental-health benefits of mindfulness in different patient categories as well as in healthy adults and children. Studies have shown a positive relationship between trait mindfulness (which can be cultivated through the practice of mindfulness-based interventions) and psychological health. The practice of mindfulness appears to provide therapeutic benefits to people with psychiatric disorders, including moderate benefits to those with psychosis. Studies also indicate that rumination and worry contribute to a variety of mental disorders, and that mindfulness-based interventions can enhance trait mindfulness and reduce both rumination and worry. Further, the practice of mindfulness may be a preventive strategy to halt the development of mental-health problems. Mindfulness practices have been said to enable individuals to respond more effectively to stressful situations by helping them strike the balance between over-identification and suppression of their emotional experiences by finding the middle point which is recognition and acceptance.

Evidence suggests that engaging in mindfulness meditation may influence physical health. For example, the psychological habit of repeatedly dwelling on stressful thoughts appears to intensify the physiological effects of the stressor (as a result of the continual activation of the sympathetic nervous system and the hypothalamus-pituitary-adrenal axis) with the potential to lead to physical health related clinical manifestations.

Studies indicate that mindfulness meditation, which brings about reductions in rumination, may alter these biological clinical pathways. Further, research indicates that mindfulness may favorably influence the immune system as well as inflammation, which can consequently impact physical health, especially considering that inflammation has been linked to the development of several chronic health conditions. Other studies support these findings.

Critics have questioned both the commercialization and the over-marketing of mindfulness for health benefits—as well as emphasizing the need for more randomized controlled studies, for more methodological details in reported studies and for the use of larger sample-sizes. While mindfulness-based interventions may be effective for youth, research has not determined methods in which mindfulness could be introduced and delivered in schools.

Watching the Breath, Body-Scan and other Techniques

There are several exercises designed to develop mindfulness meditation, which may be aided by guided meditations "to get the hang of it". As forms of self-observation and interoception, these methods increase awareness of the body, so they are usually beneficial to people with low self-awareness or low awareness of their bodies or emotional state. However, it may provoke anxiety, panic attacks, depression, and dissociation, in people who are very focused on themselves, their bodies, and their emotions.

- One method is to sit in a straight-backed chair or sit cross-legged on the floor or a cushion, close one's eyes and bring attention to either the sensations of breathing in the proximity of one's nostrils or to the movements of the abdomen when breathing in and out. In this meditation practice, one does not try to control one's breathing, but attempts to simply be aware of one's natural breathing process/rhythm. When engaged in this practice, the mind will often run off to other thoughts and associations, and if this happens, one passively notices that the mind has wandered, and in an accepting, non-judgmental way, returns to focusing on breathing.
- In body-scan meditation the attention is directed at various areas of the body and noting body sensations that happen in the present moment.
- One could also focus on sounds, sensations, thoughts, feelings and actions that happen in the present. In this regard, a famous exercise, introduced by Kabat-Zinn in his MBSR program, is the mindful tasting of a raisin, in which a raisin is being tasted and eaten mindfully. By enabling reconnection with internal hunger and satiety cues, mindful eating has been suggested to be a means of maintaining healthy and conscious eating patterns.
- Other approaches include practicing yoga asanas while attending to movements and body sensations, and walking meditation.

Timings

Meditators are recommended to start with short periods of 10 minutes or so of meditation practice per day. As one practices regularly, it becomes easier to keep the attention focused on breathing. An old Zen saying suggests, "You should sit in meditation for 20 minutes every day — unless you're too busy. Then you should sit for an hour."

In Buddhist Context; Moral Precepts

In a Buddhist context the keeping of moral precepts is an essential preparatory stage for mindfulness or meditation. Vipassana also includes contemplation and reflection on phenomena as dukkha, anatta and anicca, and reflections on causation and other Buddhist teachings.

Two-Component Model

A two-component model of mindfulness based upon a consensus among clinical psychologists has been proposed as an operational and testable definition, :

- The first component involves the self-regulation of attention so that it is maintained on immediate experience, thereby allowing for increased recognition of mental events in the present moment. The second component involves adopting a particular orientation toward one's experiences in the present moment, an orientation that is characterized by curiosity, openness, and acceptance.

In this two-component model, self-regulated attention (the first component) "involves bringing awareness to current experience—observing and attending to the changing fields of "objects" (thoughts, feelings, sensations), from moment to moment – by regulating the focus of attention". Orientation to experience (the second component) involves maintaining an attitude of curiosity about objects experienced at each moment, and about where and how the mind wanders when it drifts from the selected focus of attention. Clients are asked to avoid trying to produce a particular state (e.g. relaxation), but rather to just notice each object that arises in the stream of consciousness.

The Five-Aggregate Model

An ancient model of the mind, generally known as the five-aggregate model enables one to understand the moment-to-moment manifestation of subjective conscious experience, and therefore can be a potentially useful theoretical resource to guide mindfulness interventions. This model is based upon the traditional buddhist description of the Skandhas.

The five aggregates are described as follows:

- **Material Form:** Includes both the physical body and external matter where material elements are continuously moving to and from the material body.
- **Feelings:** Can be pleasant, unpleasant or neutral.
- **Perceptions:** Represent being aware of attributes of an object (e.g. color, shape, etc.).
- **Volition:** Represents bodily, verbal, or psychological behavior.
- **Sensory Consciousness:** Refers to input from the five senses (seeing,

hearing, smelling, tasting or touch sensations) or a thought that happens to arise in the mind.

This model describes how sensory consciousness results in the generation of feelings, perception or volition, and how individuals' previously conditioned attitudes and past associations influence this generation. The five aggregates are described as constantly arising and ceasing in the present moment.

Cultivating Self-Knowledge and Wisdom

The practice of mindfulness can be utilized to gradually develop self-knowledge and wisdom. In this regard, Buddhist teachings provide detailed instructions on how one can carry out an inquiry into the nature of the mind, and this guidance can help one to make sense of one's subjective experience. This could include understanding what the "present moment" is, how various thoughts, etc., arise following input from the senses, the conditioned nature of thoughts, and other realizations. In Buddhist teachings, ultimate wisdom refers to gaining deep insight into all phenomena or "seeing things as they are."

Applications

According to Jon Kabat-Zinn the practice of mindfulness may be beneficial to many people in Western society who might be unwilling to adopt Buddhist traditions or vocabulary. Western researchers and clinicians who have introduced mindfulness practice into mental health treatment programs usually teach these skills independently of the religious and cultural traditions of their origins. Programs based on MBSR and similar models have been widely adopted in schools, prisons, hospitals, veterans centers, and other environments.

Mindfulness-based stress reduction (MBSR) is a mindfulness-based program developed by Jon Kabat-Zinn at the University of Massachusetts Medical Center, which uses a combination of mindfulness meditation, body awareness, and yoga to help people become more mindful. While MBSR has its roots in spiritual teachings, the program itself is secular.

Mindfulness-based cognitive therapy (MBCT) is a psychological therapy designed to aid in preventing the relapse of depression, specifically in individuals with Major depressive disorder (MDD). It uses traditional cognitive behavioral therapy (CBT) methods and adds in newer psychological strategies such as mindfulness and mindfulness meditation. Cognitive methods can include educating the participant about depression. Mindfulness and mindfulness meditation focus on becoming aware of all incoming thoughts and feelings and accepting them, but not attaching or reacting to them.

Like CBT, MBCT functions on the theory that when individuals who have historically had depression become distressed, they return to automatic cognitive processes that can trigger a depressive episode. The goal of MBCT is to interrupt these automatic processes and teach the participants to focus less on reacting to incoming stimuli, and instead accepting and observing them without judgment. This mindfulness practice allows the participant to notice when automatic processes are occurring and to alter their reaction to be more of a reflection. Research supports the effects of MBCT in people who have been depressed three or more times and demonstrates reduced relapse rates by 50%.

Mindfulness-based pain management (MBPM) is a mindfulness-based intervention (MBI) providing specific applications for people living with chronic pain and illness. Adapting the core concepts and practices of mindfulness-based stress reduction (MBSR) and mindfulness-based cognitive therapy (MBCT), MBPM includes a distinctive emphasis on the practice of 'loving-kindness', and has been seen as sensitive to concerns about removing mindfulness teaching from its original ethical framework. It was developed by Vidyamala Burch and is delivered through the programs of Breathworks. It has been subject to a range of clinical studies demonstrating its effectiveness.

Acceptance and commitment therapy or (ACT) (typically pronounced as the word "act") is a form of clinical behavior analysis (CBA) used in psychotherapy. It is a psychological intervention that uses acceptance and mindfulness strategies mixed in different ways with commitment and behavior-change strategies, to increase psychological flexibility. The approach was originally called *comprehensive distancing*. It was developed in the late 1980s by Steven C. Hayes, Kelly G. Wilson, and Kirk Strosahl.

Mindfulness is a "core" exercise used in dialectical behavior therapy (DBT), a psychosocial treatment Marsha M. Linehan developed for treating people with borderline personality disorder. DBT is dialectic, says Linehan, in the sense of "the reconciliation of opposites in a continual process of synthesis." As a practitioner of Buddhist meditation techniques, Linehan says:

- This emphasis in DBT on a balance of acceptance and change owes much to my experiences in studying meditation and Eastern spirituality. The DBT tenets of observing, mindfulness, and avoidance of judgment are all derived from the study and practice of Zen meditations.

Mode deactivation therapy (MDT) is a treatment methodology that is derived from the principles of cognitive-behavioral therapy and incorporates elements of Acceptance and commitment therapy, Dialectical behavior therapy, and mindfulness techniques. Mindfulness techniques such as simple breathing exercises are applied to assist the client in awareness and non-judgmental acceptance of unpleasant and distressing thoughts and feelings as they occur in the present moment. Mode Deactivation Therapy was developed and is established as an effective treatment for adolescents with problem behaviors and complex trauma-related psychological problems, according to recent publications by Jack A. Apsche and Joan Swart.

The Japanese psychiatrist Shoma Morita, who trained in Zen meditation, developed Morita therapy upon principles of mindfulness and non-attachment.

Internal Family Systems Model (IFS), developed by Richard C. Schwartz, emphasizes the importance of both therapist and client engaging in therapy from the Self, which is the IFS term for one's "spiritual center". The Self is curious about whatever arises in one's present experience and open and accepting toward all manifestations.

Mindfulness relaxation uses breathing methods, guided imagery, and other practices to relax the body and mind and help reduce stress.

In 2012 Congressman Tim Ryan of Ohio published *A Mindful Nation*, and received a $1 million federal grant to teach mindfulness in schools in his home district.

Mindful Kids Miami is a tax-exempt, 501 (c)(3), non-profit corporation established in 2011 dedicated to making age-appropriate mindfulness training available to school children in Miami-Dade County public and private schools. This is primarily accomplished by training educators and other childcare providers to incorporate mindfulness practices in the children's daily activities.

In 2000, The Inner Kids Program, a mindfulness-based program developed for children, was introduced into public and private school curricula in the greater Los Angeles area.

MindUP, a classroom-based program spearheaded by Goldie Hawn's Hawn Foundation, teaches students to self-regulate behavior and mindfully engage in focused concentration required for academic success. For the last decade, MindUP has trained teachers in over 1,000 schools in cities from Arizona to Washington.

The Holistic Life Foundation, a non-profit organization that created an in-school mindfulness program called Mindful Moment, is currently serving almost 350 students daily at Robert W. Coleman Elementary School and approximately 1300 students at Patterson Park High School in Baltimore, Maryland. At Patterson High School, the Mindful Moment program engages the school's faculty along with the students during a 15-minute mindfulness practice at the beginning and end of each school day.

Mindful Life Project, a non-profit 501(c)3 based out of Richmond, California, teaches mindfulness to elementary school students in underserved schools in the South Richmond school district. Utilizing curriculum, "Rise-Up" is a regular school day intervention program serving 430 students weekly, while "Mindful Community" is currently implemented at six South Richmond partner schools. These in-school mindfulness programs have been endorsed by Richmond Mayor Gayle McLaughlin, who has recommended additional funding to expand the program in order to serve all Richmond youth.

Mindfulness practices are becoming more common within educational institutions including Elementary and Secondary schools. This has been referred to as part of a 'contemplative turn' in education that has emerged since the turn of the millennium. The applications of mindfulness in schools are aimed at calming and relaxation of students as well as for students and educators to build compassion and empathy for others. An additional benefit to Mindfulness in education is for the practice to reduce anxiety and stress in students. Based on a broad meta-analytical review, scholars said that the application of mindfulness practice enhances the goals of education in the 21st century, which include adapting to a rapidly changing world and being a caring and committed citizen. Within educational systems, the application of mindfulness practices shows an improvement of students' attention and focus, emotional regulation, creativity, and problem solving skills. As discussed by Ergas and Todd, the development of this field since the turn of the millennium has brought diverse possibilities as well as complexities, given the origins of mindfulness within Buddhism and the processes of its secularization and measurement based on science.

Renshaw and Cook state, "As scientific interest in the utility of Mindfulness-Based Intervention (MBI) in schools grew steadily, popular interest in mindfulness in schools seemed to grow exponentially". Despite research on mindfulness being comparatively unexamined, especially with young students, the practice has seen a spike in use within the educational arena. "A relatively recent addition to discourse around preventing school expulsion and failure, mindfulness is gaining popularity for its potential to improve students' social, emotional, behavioral, and learning-related cognitive control, thereby improving academic outcomes". Researchers and educators are interested in how mindfulness can provide optimal conditions for a students' personal development and academic success. Current research on mindfulness in education is limited but can provide insight into the potential benefits for students, and areas of improvement for future studies.

Mindfulness in the classroom is being touted as a promising new intervention tool for young students. According to Choudhury and Moses, "Although still marginal and in some cases controversial, secular programs of mindfulness have been implemented with ambitious goals of improving attentional focus of pupils, social-emotional learning in "at-risk" children and youth, not least, to intervene in problems of poverty and incarceration". Emerging research is concerned with studying teachers and programs using mindfulness practices with students and is discovering tension arising from the moral reframing of eastern practices in western school settings. As cited by Renshaw and Cook, "Unlike most other approaches to contemporary school-based intervention, which are squarely grounded in behavioral, cognitive-behavioral, and ecological systems theories, MBIs have their origins in Eastern religious traditions". Some school administrators are concerned about implementing such practices, and parents have been reported to take their children out of mindfulness programs because of their personal religious beliefs. Yet, MBIs continue to be accepted by the mainstream in both primary and secondary schools because, "Mindfulness practices, particularly in relation to children who might otherwise be considered broken or unredeemable, fill a critical niche – one that allows its advocates to imagine a world where people can change, become more compassionate, resilient, reflective, and aware; a world with a viable future". As mindfulness in education continues to develop, ethical consequences will remain a controversial issue because the generic description for the "benefits" and "results" of MBIs are largely concerned with individual and inward-focused achievement, rather than the original Buddhist ideal of global human connection.

Available research reveals a relationship between mindfulness and attention. Semple, Lee, Rosa, & Miller say, "Anxiety can impair attention and promote emotionally reactive behaviors that interfere with the development of good study skills, so it seems reasonable that increased mindfulness would be associated with less anxiety". They conducted a randomized trial of Mindfulness-Based Cognitive Therapy for Children (MBCT-C) that found promise in managing anxiety for elementary school-aged children, and suggests that those who completed the program displayed fewer attention problems. In addition, Flook shows how an eight-week mindfulness awareness program was evaluated in a random and controlled school setting and measured the effects of awareness practices on executive functions in elementary school children.

Their findings concluded, "Participation in the mindfulness awareness program was associated with improvements in behavioral regulation, metacognition, and overall executive functions". In the study by Flook, parents and teachers completed questionnaires which propose that participation in mindfulness programs is associated with improvements in child behavioral regulation.

These perspectives are a valuable source of data given that caregivers and educators interact with the children daily and across a variety of settings. According to Eklund, Omalley, and Meyer, "School-based practitioners should find promise in the evidence supporting mindfulness-based practices with children, parents, and educators". Lastly, a third study by Zenner, Herrnleben-Kurz, and Walach concluded, "Analysis suggest that mindfulness-based interventions for children and youths are able to increase cognitive capacity of attending and learning by nearly one standard deviation and yield". Application of Mindfulness-Based Interventions continue to increase in popularity and practice.

Mindfulness-Based Interventions are rising across western culture, but its effectiveness in school programs is still being determined. Research contends, "Mindfulness-based approaches for adults are effective at enhancing mental health, but few controlled trials have evaluated their effectiveness among young people". Although much of the available studies find a high number of mindfulness acceptability among students and teachers, more research needs to be conducted on its effects on well-being and mental health for students. In a firmly controlled experiment, Johnson, Burke, Brinkman, and Wade evaluated "the impact of an existing and widely available school-based mindfulness program". According to their research, "no improvements were demonstrated on any outcome measured either immediately post-intervention or at three-month follow-up". Many questions remain on which practices best implement effective and reliable mindfulness programs at schools, and further research is needed to identify the optimal methods and measurement tools for mindfulness in education.

Mindfulness training appears to be getting popular in the business world, and many large corporations have been incorporating mindfulness practices into their culture. For example, companies such as Google, Apple, Procter & Gamble, General Mills, Mayo Clinic, and the U.S. Army offer mindfulness coaching, meditation breaks and other resources to their employees to improve workplace functioning.

The introduction of mindfulness in corporate settings still remains in early stages and its potential long-term impact requires further assessment. Mindfulness has been found to result in better employee well-being, lower levels of frustration, lower absenteeism and burnout as well as an improved overall work environment.

Legal and law enforcement organizations are also showing interest in mindfulness:

- Harvard Law School's Program on Negotiation hosted a workshop on "Mindfulness in the Law & Alternative Dispute Resolution."
- Many law firms offer mindfulness classes.

Mindfulness has been taught in prisons, reducing hostility and mood disturbance among inmates, and improving their self-esteem. Additional studies indicate that mindfulness interventions can result in significant reductions in anger, reductions in substance use, increased relaxation capacity, self-regulation and optimism.

Many government organizations offer mindfulness training. Coping Strategies is an example of a program utilized by United States Armed Forces personnel. The British Parliament organized a mindfulness-session for its members in 2014, led by Ruby Wax.

Effects and Efficacy of Mindfulness Practice

Mindfulness has gained increasing empirical attention since 1970 and has been studied often as an intervention for stress reduction. Meta analyses indicate its beneficial effects for healthy adults, for adolescents and children, as well as for different health-related outcomes including weight management, psychiatric conditions, heart disease, sleep disorders, cancer care, adult autism treatment, multiple sclerosis, and other health-related conditions. An often-cited meta-analysis on meditation research published in JAMA in 2014, found insufficient evidence of any effect of meditation programs on positive mood, attention, substance use, eating habits, sleep, and weight, but found that there is moderate evidence that meditation reduces anxiety, depression, and pain. However, this study included a highly heterogeneous group of meditation styles (i.e., it did not focus exclusively on mindfulness meditation), which is a significant limitation of this study. Additionally, while mindfulness is well known to have positive psychological effect among individuals diagnosed with various types of cancers, the evidence is unclear regarding its effectiveness in men with prostate cancer.

Thousands of studies on meditation have been conducted, though the methodological quality of some of the studies is poor. Recent reviews have described many of these issues. Nonetheless, mindfulness meditation is a popular subject for research, and many present potential benefits for a wide array of conditions and outcomes. For example, the practice of mindfulness has also been used to improve athletic performance, as a beneficial intervention for children with special needs and their caregivers, as a viable treatment option for people with insomnia an effective intervention for healthy aging, as a strategy for managing dermatological conditions and as a useful intervention during early pregnancy. Recent studies have also demonstrated that mindfulness meditation significantly attenuates physical pain through multiple, unique mechanisms. Meditation also may allow one to modulate pain. When exposed to pain from heating, the brain scans of the mindfulness meditation participants (by use of functional magnetic resonance imaging) showed their brains notice the pain equally, however it does not get converted to a perceived pain signal. As such they experienced up to 40–50% less pain. Research has also investigated mindful movements and mindful exercises for different patient populations.

Neurological Studies

Research studies have also focused on the effects of mindfulness on the brain using neuroimaging techniques, physiological measures and behavioral tests. Research on the neural perspective of how mindfulness meditation works suggests that it exerts its effects in components of attention regulation, body awareness and emotional regulation. When considering aspects such as sense of responsibility, authenticity, compassion, self-acceptance and character, studies have shown that mindfulness meditation contributes to a more coherent and healthy sense of self and identity. Neuroimaging techniques suggest that mindfulness practices such as mindfulness meditation are associated with "changes in the anterior cingulate cortex, insula, temporo-parietal junction, fronto-limbic network and default mode network structures." Further, mindfulness meditation may prevent or delay the onset of mild cognitive impairment and Alzheimer's disease. Additionally, mindfulness-induced emotional and behavioral changes have been found to be related to functional and structural changes in the brain. It has also been suggested that the default mode network of the brain can be used as a potential biomarker for monitoring the therapeutic benefits of meditation. Recent research also suggest that the practice of mindfulness could influence genetic expression leading to a reduced risk of inflammation-related diseases and favourable changes in biomarkers.

Grey matter concentrations in brain regions that regulate emotion, self-referential processing, learning and memory processes have shown changes in density following MBSR.

Additionally, MBSR practice has been associated with improvement of the immune system which could explain the correlation between stress reduction and increased quality of life. Part of these changes are a result of the thickening of the prefrontal cortex (executive functioning) and hippocampus (learning and memorisation ability), the shrinking of the amygdala (emotion and stress response) and the strengthening of the connections between brain cells. Long-term meditators have larger amounts of gyrification ("folding" of the cortex, which may allow the brain to process information faster) than people who do not meditate. Further, a direct correlation was found between the amount of gyrification and the number of meditation years, possibly providing further proof of the brain's neuroplasticity, or ability to adapt to environmental changes.

Associations of Mindfulness with other Variables

Mindfulness (as a trait, distinguished from mindfulness *practice*) has been linked to many outcomes. In an overview, Keng, Smoski, and Robins summarize: "Trait mindfulness has been associated with higher levels of life satisfaction, agreeableness, conscientiousness, vitality, self esteem, empathy, sense of autonomy, competence, optimism, and pleasant affect.

A 2020 study found links between dispositional mindfulness and prosocial behavior. Studies have also demonstrated significant negative correlations between mindfulness and depression, neuroticism, absentmindedness, dissociation, rumination, cognitive reactivity, social anxiety, difficulties in emotion regulation, experiential avoidance, alexithymia, intensity of delusional experience in the context of psychosis, and general psychological symptoms."

Effects on Mindfulness

The mechanisms that make people less or more mindful have been researched less than the effects of mindfulness programmes, so little is known about which components of mindfulness practice are relevant for promoting mindfulness. For example, meta-analyses have shown that mindfulness practice does increase mindfulness when compared to active control groups. This may be because we do not know how to measure mindfulness. It could also be that mindfulness is dose-dependent and increases with more experience. To counter that, Bergomi et al. found that "results provide evidence for the associations between self-reported mindfulness and meditation practice and suggest that mindfulness is particularly associated with continued practice in the present, rather than with accumulated practice over years."

Some research into other mechanisms has been done. One study conceptualized such mechanisms in terms of competition for attention. In a test of that framework, mindfulness was found to be associated (as predicted) with having an activated intention to be mindful, with feeling good, and with not being hurried or very busy. Regarding the relationship between feeling good and being mindful, a different study found that causality probably works both ways: feeling good increases mindfulness, and mindfulness increases feeling good.

One theory suggests an additional mechanism termed as *reperceiving*. Reperceiving is the beneficial effect that comes after the process of being mindful after all the intention, attention, and attitude has been experienced. Through reperceiving there is a shift in perspective. Reperceiving permits disassociation from thoughts, emotions, and physical sensations, and allows one to exist with them instead of being defined by them.

5.1.4 Technological Advancements in Cognitive Enhancement

Neurostimulation Techniques

Neurostimulation techniques, such as transcranial magnetic stimulation (TMS) and transcranial direct current stimulation (tDCS), modulate neural activity to enhance cognitive functions. TMS delivers magnetic pulses to specific brain regions implicated in memory and learning, while tDCS uses low-intensity electrical currents to alter neuronal excitability. These non-invasive techniques have demonstrated potential in enhancing working memory, attentional control, and decision-making skills.

Wearable Devices and Cognitive Assistants

Wearable devices equipped with biosensors and AI algorithms monitor physiological parameters, detect cognitive patterns, and provide real-time feedback to optimize cognitive performance. Cognitive assistants powered by AI offer personalized recommendations, reminders, and adaptive learning programs tailored to individual cognitive strengths and weaknesses. These technologies support cognitive rehabilitation, mental health management, and everyday cognitive enhancement strategies.

5.1.5 Ethical Considerations in Cognitive Enhancement

Privacy and Data Security

The integration of technology in cognitive enhancement raises concerns about privacy protection and data security. Personalized cognitive training programs and neurostimulation devices collect sensitive data related to brain function and behavior. Safeguarding user privacy, obtaining informed consent, and implementing robust cybersecurity measures are essential to prevent unauthorized access and misuse of cognitive data.

Equity and Access

Ensuring equitable access to cognitive enhancement technologies is critical to mitigate disparities based on socioeconomic status, geographic location, or disability. Addressing affordability, digital literacy, and cultural factors promotes inclusive participation in cognitive enhancement interventions and prevents exacerbation of existing inequalities. Ethical guidelines and regulatory frameworks should prioritize accessibility and fair distribution of cognitive enhancement resources.

Autonomy and Enhancement Risks

Ethical debates surrounding cognitive enhancement focus on autonomy, authenticity, and potential risks of unintended consequences. Enhancing cognitive abilities through pharmacological interventions, neurostimulation, or genetic enhancements raises ethical dilemmas regarding fairness in competitive environments, coercion to enhance performance, and long-term implications for personal identity and societal norms.

5.1.6 Future Directions in Cognitive Enhancement

Precision Medicine Approaches

Advancements in genetics, neuroimaging, and biomarker research enable precision medicine approaches to cognitive enhancement. Personalized interventions based on individual genetic profiles, neurophysiological characteristics, and cognitive assessments optimize treatment efficacy and minimize adverse effects. Targeted therapies for neurodegenerative disorders and age-related cognitive decline are under investigation to preserve cognitive function and enhance quality of life.

Brain-Computer Interfaces (BCIs) and Neural Implants

Emerging technologies, such as BCIs and neural implants, offer direct interfaces between the brain and external devices to restore or augment cognitive functions. BCIs decode neural signals to control prosthetic devices, communicate thoughts, or enhance attention and memory capacities. Research focuses on improving BCI accuracy, miniaturization, and biocompatibility for safe and effective integration into clinical and everyday applications.

Ethical Guidelines and Responsible Innovation

As cognitive enhancement technologies evolve, establishing ethical guidelines, regulatory oversight, and public engagement initiatives is crucial for responsible innovation. Collaboration among scientists, policymakers, ethicists, and stakeholders promotes ethical decision-making, transparency in research practices, and inclusive societal dialogue. Ethical frameworks should prioritize safety, autonomy, privacy rights, and equitable access to ensure ethical adoption and societal acceptance of cognitive enhancement advancements.

5.2 COGNITIVE COMPUTING IN HUMAN-MACHINE COLLABORATION

Cognitive computing represents a transformative approach to artificial intelligence (AI) that aims to mimic human thought processes and enhance decision-making capabilities in machines. Cognitive computing represents a paradigm shift in human-machine collaboration, empowering organizations and individuals to harness AI-driven capabilities for enhanced decision-making, productivity, and innovation. From healthcare diagnostics and financial analytics to customer service interactions and educational technologies, cognitive computing systems leverage advanced AI algorithms, big data analytics, and cloud computing infrastructure to deliver personalized experiences and optimize outcomes. Addressing ethical considerations, including bias mitigation, privacy protection, and algorithmic transparency, is crucial for fostering trust, accountability, and societal acceptance of cognitive computing technologies. Looking ahead, advancing research, ethical guidelines, and human-centered design principles will shape the future landscape of cognitive computing, promoting responsible innovation and equitable benefits for global communities.

5.2.1 Understanding Cognitive Computing

Definition and Objectives

Cognitive computing refers to systems that simulate human cognitive abilities, such as reasoning, learning, problem-solving, and decision-making, using AI algorithms and data analytics. Unlike traditional AI, which relies on predefined rules and algorithms, cognitive computing systems leverage machine learning, natural language processing (NLP), and pattern recognition to process vast amounts of unstructured data and derive meaningful insights.

Key Components

Key components of cognitive computing systems include:

- **Machine Learning:** Algorithms that enable machines to learn from data, identify patterns, and make predictions without explicit programming.
- **Natural Language Processing (NLP):** Techniques that enable machines to understand and generate human language, facilitating interaction and communication.
- **Pattern Recognition:** Algorithms that analyze and interpret complex patterns in data, enabling cognitive systems to recognize images, speech, and behavioral patterns.

5.2.2 Applications of Cognitive Computing

Healthcare and Medicine

Cognitive computing has transformative applications in healthcare, enabling personalized diagnostics, treatment recommendations, and drug discovery. AI-powered systems analyze medical imaging data, electronic health records (EHRs), and genomic data to assist healthcare professionals in diagnosing diseases, predicting patient outcomes, and developing tailored treatment plans. Virtual health assistants powered by cognitive computing provide patient education, monitor chronic conditions, and improve healthcare delivery efficiency.

Finance and Business Analytics

In finance and business analytics, cognitive computing enhances decision-making processes by analyzing market trends, customer behavior, and financial data in real-time. AI-driven systems predict market fluctuations, recommend investment strategies, and optimize operational efficiency. Natural language processing capabilities enable cognitive systems to process and interpret financial news, reports, and regulatory documents, supporting risk management and compliance activities.

Customer Service and Interaction

Cognitive computing improves customer service and interaction through chatbots, virtual assistants, and voice recognition systems. AI-powered chatbots leverage natural language understanding to provide personalized responses, resolve customer inquiries, and streamline customer support processes. Virtual assistants enhance user experiences by anticipating user needs, scheduling appointments, and facilitating seamless interactions across digital platforms.

5.2.3 Technologies Underpinning Cognitive Computing

Deep Learning and Neural Networks

Deep learning and neural networks are foundational technologies in cognitive computing, enabling machines to process and analyze complex data sets. Deep neural networks mimic the structure and function of the human brain, learning hierarchical representations of data and extracting meaningful features for decision-making. Advances in deep learning algorithms, such as convolutional neural networks (CNNs) and recurrent neural networks (RNNs), drive innovations in computer vision, natural language understanding, and autonomous systems.

Big Data Analytics

Cognitive computing harnesses big data analytics to derive actionable insights from large volumes of structured and unstructured data. Advanced data processing techniques, including data mining, predictive analytics, and anomaly detection, enable cognitive systems to uncover hidden patterns, correlations, and trends that inform decision-making and strategy formulation. Real-time data analytics empower organizations to respond quickly to changing market conditions and customer preferences.

Cloud Computing and Edge Computing

Cloud computing and edge computing infrastructure support cognitive computing applications by providing scalable computing resources and storage capabilities. Cloud-based cognitive services offer on-demand access to AI models, data repositories, and computational resources for developing and deploying cognitive applications. Edge computing enhances real-time processing and responsiveness by decentralizing computation and data storage closer to end-users, supporting applications in IoT devices, autonomous vehicles, and smart cities.

5.2.4 Ethical Considerations in Cognitive Computing

Bias and Fairness

Ethical concerns in cognitive computing include bias in AI algorithms, which can perpetuate social inequalities and discrimination. Biased training data, algorithmic biases, and lack of diversity in AI development teams contribute to biased outcomes in decision-making processes. Addressing bias requires transparent data collection practices, diverse representation in AI development, and algorithmic fairness assessments to mitigate unintended consequences and promote equitable outcomes.

Privacy and Data Protection

Cognitive computing raises privacy concerns regarding the collection, storage, and use of personal data for AI-driven applications. Protecting sensitive information, ensuring data security, and obtaining informed consent from users are essential for safeguarding privacy rights in cognitive computing systems. Compliance with data protection regulations, such as GDPR and CCPA, establishes guidelines for responsible data management and user privacy rights.

Accountability and Transparency

Ensuring accountability and transparency in cognitive computing involves identifying responsible parties for AI decisions, disclosing AI capabilities and limitations, and providing explanations for algorithmic outcomes. Establishing ethical guidelines, regulatory frameworks, and industry standards promotes responsible AI deployment and fosters trust among stakeholders. Transparent communication about AI systems' capabilities, data usage, and potential risks enhances public understanding and acceptance of cognitive computing technologies.

5.2.5 Societal Impact of Cognitive Computing

Employment and Workforce Dynamics

The adoption of cognitive computing technologies reshapes employment and workforce dynamics by automating routine tasks, augmenting human capabilities, and creating new job roles. AI-driven automation improves productivity in industries such as manufacturing, logistics, and customer service, while generating demand for skills in AI development, data science, and human-machine collaboration. Upskilling and reskilling initiatives prepare the workforce for roles that require creativity, problem-solving, and ethical decision-making in AI-enabled environments.

Education and Lifelong Learning

Cognitive computing transforms education and lifelong learning through personalized learning platforms, adaptive tutoring systems, and AI-driven educational content. AI-powered tools analyze student performance data, identify learning gaps, and tailor instructional materials to individual learning styles and abilities. Virtual reality (VR) and augmented reality (AR) applications enhance immersive learning experiences, simulating real-world scenarios and promoting interactive engagement in educational settings.

Healthcare Accessibility and Patient Outcomes

In healthcare, cognitive computing improves accessibility to medical expertise, enhances diagnostic accuracy, and optimizes treatment outcomes. AI-driven diagnostic tools assist healthcare providers in detecting diseases at earlier stages, predicting patient responses to treatments, and recommending personalized interventions. Remote monitoring and telehealth services powered by cognitive computing expand access to healthcare services, particularly in underserved communities and rural areas.

5.2.6 Future Directions in Cognitive Computing

Explainable AI and Interpretability

Future advancements in cognitive computing will focus on developing explainable AI models and enhancing algorithmic interpretability. Explainable AI techniques enable AI systems to provide transparent explanations for decision-making processes, making AI-driven recommendations understandable and trustworthy for users. Enhancing interpretability fosters accountability, facilitates regulatory compliance, and promotes ethical adoption of cognitive computing technologies across diverse applications.

Human-Centered AI Design

Human-centered AI design principles prioritize user-centric considerations, ethical values, and societal impact in AI development and deployment. Collaborative approaches involving interdisciplinary teams, end-user feedback, and stakeholder engagement promote inclusive design practices that prioritize human well-being, fairness, and safety in cognitive computing systems. Designing AI systems with empathy, cultural sensitivity, and accessibility considerations enhances usability, acceptance, and effectiveness in diverse user environments.

Ethical Guidelines and Governance Frameworks

Establishing ethical guidelines, governance frameworks, and regulatory standards is essential to ensure responsible innovation and mitigate risks associated with cognitive computing technologies.

Multistakeholder collaboration among policymakers, industry leaders, academics, and civil society organizations fosters consensus on ethical principles, data governance practices, and AI accountability mechanisms. Ethical AI certifications and audit procedures promote compliance with ethical standards and encourage continuous improvement in AI governance.

5.3 ETHICAL CONSIDERATIONS IN NEUROENHANCEMENT

Neuroenhancement, the use of substances or technologies to improve cognitive functions beyond normal levels, raises profound ethical questions regarding safety, fairness, autonomy, and societal impact. Ethical considerations in neuroenhancement encompass autonomy, safety, fairness, and societal impact, shaping ethical discourse, policy development, and responsible use of cognitive enhancement technologies. From pharmacological substances and brain stimulation techniques to cognitive training programs and ethical guidelines, navigating ethical dilemmas in neuroenhancement requires interdisciplinary collaboration, informed consent, and regulatory oversight. Addressing ethical concerns promotes ethical literacy, fosters public trust, and advances neuroenhancement research toward enhancing cognitive functions while upholding ethical principles, human rights, and societal well-being.

5.3.1 Understanding Neuroenhancement

Definition and Scope

Neuroenhancement refers to the augmentation of cognitive functions, such as memory, attention, creativity, and decision-making, through pharmacological substances, brain stimulation techniques, or cognitive training methods.

The goal of neuroenhancement is to optimize brain performance and enhance mental capabilities beyond natural abilities, potentially improving academic, professional, or personal achievements.

Pharmacological Substances

Pharmacological neuroenhancers, such as stimulants (e.g., methylphenidate, amphetamines) and cognitive enhancers (e.g., modafinil, donepezil), alter neurotransmitter activity in the brain to enhance cognitive functions.

These substances are commonly used off-label for their effects on focus, alertness, and cognitive performance, particularly among students, professionals, and individuals seeking cognitive enhancement.

Brain Stimulation Techniques

Non-invasive brain stimulation techniques, including transcranial magnetic stimulation (TMS) and transcranial direct current stimulation (tDCS), modulate neuronal activity to improve cognitive functions.

TMS delivers magnetic pulses to specific brain regions implicated in memory and learning, while tDCS uses low-intensity electrical currents to enhance neural excitability and promote cognitive performance.

Cognitive Training and Behavioral Interventions

Cognitive training programs, such as brain training apps and memory exercises, aim to improve cognitive skills through repetitive tasks, problem-solving exercises, and memory enhancement techniques.

Behavioral interventions, including mindfulness practices and cognitive-behavioral therapy (CBT), promote cognitive resilience, emotional regulation, and adaptive coping strategies to optimize mental well-being and cognitive functioning.

5.3.2 Ethical Considerations in Neuroenhancement

Autonomy and Informed Consent

Respecting individuals' autonomy and ensuring informed consent are essential ethical principles in neuroenhancement. Individuals should have the right to make informed decisions about using neuroenhancement methods, including understanding potential risks, benefits, and alternatives. Informed consent requires transparency in disclosing information about neuroenhancement substances, side effects, and long-term implications for personal health and well-being.

Safety and Risk-Benefit Analysis

Assessing the safety and efficacy of neuroenhancement methods involves balancing potential benefits with risks to individual health and safety. Pharmacological substances may have adverse effects, including addiction, cardiovascular complications, and psychological dependence. Brain stimulation techniques carry risks of neurological side effects and require careful monitoring by trained professionals to ensure safe and effective application.

Fairness and Equity

Promoting fairness and equity in access to neuroenhancement raises concerns about socioeconomic disparities, academic pressure, and competitive advantages in educational and professional settings. Access to neuroenhancement substances and technologies should not exacerbate existing inequalities or create unfair advantages based on financial resources, access to healthcare, or social privilege. Addressing equity requires policies that promote equal opportunities, support diverse learning styles, and mitigate pressures to use neuroenhancement for competitive gain.

Enhancement vs. Treatment

Distinguishing between neuroenhancement for cognitive optimization and medical treatment for cognitive deficits is crucial for ethical decision-making. Neuroenhancement aims to enhance cognitive functions in healthy individuals without medical necessity, whereas medical treatments address diagnosed conditions, such as attention deficit hyperactivity disorder (ADHD), Alzheimer's disease, or cognitive impairment due to neurological disorders. Ethical guidelines should ensure appropriate use of neuroenhancement methods, prioritizing medical need, and clinical supervision for individuals with cognitive impairments.

5.3.3 Societal Impact of Neuroenhancement

Education and Academic Integrity

Neuroenhancement in educational settings raises ethical concerns about academic integrity, fairness in assessment, and pressures to achieve academic success. Students may use cognitive enhancers to cope with academic demands, enhance concentration, and improve exam performance, potentially compromising academic honesty and undermining educational values. Promoting ethical conduct, supporting mental health resources, and fostering a culture of academic integrity are essential in mitigating pressures for neuroenhancement in academic environments.

Workplace Performance and Professional Ethics

In the workplace, neuroenhancement technologies may influence professional ethics, productivity expectations, and ethical decision-making among employees. Enhancing cognitive functions through pharmacological substances or brain stimulation techniques can impact job performance, decision-making processes, and interpersonal interactions. Ethical guidelines in organizational settings should address the responsible use of neuroenhancement methods, respect for employee autonomy, and policies that promote employee well-being and ethical conduct.

Public Perception and Stigma

Public perception of neuroenhancement influences societal acceptance, ethical discourse, and regulatory policies governing cognitive enhancement practices. Stigma surrounding cognitive enhancers, concerns about fairness, and ethical implications shape public attitudes toward neuroenhancement technologies. Educating the public, fostering informed discussions, and promoting ethical literacy in neuroscience ethics enhance understanding of neuroenhancement issues and inform public policy debates.

5.3.4 Regulatory Challenges and Ethical Guidelines

Regulatory Oversight

Regulating neuroenhancement presents challenges related to defining legal frameworks, monitoring safety and efficacy, and addressing ethical concerns in diverse cultural and regulatory contexts. National regulatory agencies, such as the FDA in the United States and the EMA in Europe, oversee approval processes for pharmaceutical neuroenhancers, ensuring compliance with safety standards and ethical guidelines. International collaboration and regulatory harmonization efforts promote consistency in evaluating neuroenhancement technologies and protecting public health.

Ethical Guidelines for Responsible Use

Developing ethical guidelines for responsible neuroenhancement involves interdisciplinary collaboration among ethicists, neuroscientists, policymakers, and stakeholders. Ethical principles, including beneficence, non-maleficence, justice, and respect for autonomy, guide ethical decision-making in neuroenhancement research, practice, and policy development. Promoting transparency, accountability, and public engagement fosters ethical awareness, informed consent, and ethical responsibility among users and providers of neuroenhancement technologies.

5.3.5 Future Directions in Ethical Neuroenhancement Research

Neuroethical Education and Awareness

Advancing neuroethical education and awareness initiatives promotes ethical literacy, critical thinking, and responsible decision-making regarding neuroenhancement. Integrating neuroscience ethics into educational curricula, professional training programs, and public outreach campaigns enhances understanding of ethical issues, fosters ethical competence, and empowers individuals to navigate ethical dilemmas in neuroenhancement contexts.

Ethical Innovation and Human-Centered Design

Human-centered design principles prioritize ethical values, user needs, and societal impact in developing neuroenhancement technologies. Collaborative approaches involving diverse stakeholders, end-user perspectives, and ethical impact assessments ensure ethical innovation, mitigate risks, and promote responsible use of neuroenhancement methods. Designing AI-driven cognitive enhancement tools with ethical guidelines, user privacy protections, and regulatory compliance supports ethical decision-making and enhances public trust in neuroenhancement technologies.

Policy Recommendations and Governance Frameworks

Policy recommendations for neuroenhancement governance include establishing ethical guidelines, monitoring neuroenhancement practices, and promoting evidence-based research on cognitive enhancement methods. Multistakeholder engagement, policy dialogue, and international cooperation facilitate consensus-building, regulatory harmonization, and responsible oversight of neuroenhancement technologies. Strengthening ethical governance frameworks safeguards public health, protects individual rights, and promotes ethical standards in advancing neuroenhancement research, practice, and policy implementation.

CHAPTER 6

EXOSKELETONS AND WEARABLE TECHNOLOGIES

6.1 POWERED EXOSKELETONS FOR MOBILITY

Powered exoskeletons are wearable robotic devices designed to augment human mobility, providing enhanced strength, endurance, and support. These advanced systems integrate mechanical and electronic components to assist or amplify human movement, benefiting individuals with mobility impairments and offering potential applications in various fields such as healthcare, military, and industrial sectors.

Powered exoskeletons represent a significant technological advancement in enhancing human mobility, offering transformative benefits for individuals with mobility impairments and various industries. While challenges related to technical limitations, cost, and ethical considerations persist, ongoing research and innovation promise to overcome these obstacles. The integration of AI, development of soft exoskeletons, and efforts to improve accessibility will drive the future of powered exoskeletons, ensuring that they continue to empower individuals, enhance productivity, and contribute to societal well-being.

6.1.1 Development of Powered Exoskeletons

Early Innovations and Historical Context

The concept of powered exoskeletons has evolved significantly since the early 20th century. Initial designs were inspired by the need to enhance human capabilities for military and industrial purposes. Early prototypes, such as General Electric's Hardiman suit in the 1960s, aimed to amplify human strength but faced limitations due to bulky designs and inadequate power sources. Advances in materials science, robotics, and control systems over the decades have propelled the development of more sophisticated and functional exoskeletons.

Technological Advancements

Modern powered exoskeletons leverage cutting-edge technologies, including lightweight materials, compact actuators, and advanced sensors, to create more efficient and user-friendly devices. These technologies enable precise control of movement, real-time feedback, and seamless integration with the user's natural biomechanics. The incorporation of artificial intelligence (AI) and machine learning algorithms further enhances the adaptability and responsiveness of exoskeletons, optimizing their performance across diverse applications.

6.1.2 Applications of Powered Exoskeletons

Healthcare and Rehabilitation

In the healthcare sector, powered exoskeletons are transforming rehabilitation practices for individuals with mobility impairments due to spinal cord injuries, strokes, or neurological disorders. These devices facilitate gait training, muscle strengthening, and neuroplasticity by enabling repetitive and controlled movements. Exoskeleton-assisted rehabilitation programs enhance patients' functional recovery, improve walking patterns, and reduce the risk of secondary complications associated with immobility.

Industrial and Occupational Uses

Powered exoskeletons are gaining traction in industrial and occupational settings to mitigate the physical strain experienced by workers engaged in repetitive tasks or heavy lifting. By augmenting human strength and endurance, these devices reduce the risk of musculoskeletal injuries and enhance productivity. Exoskeletons designed for industrial use often feature ergonomic designs, adjustable support levels, and compatibility with various tools and equipment, promoting worker safety and efficiency.

Military and Defense Applications

The military sector has long been interested in powered exoskeletons to enhance soldiers' physical capabilities and endurance on the battlefield. These devices provide support for carrying heavy loads, reduce fatigue during prolonged missions, and improve mobility in challenging terrains. Military exoskeletons are designed to be lightweight, durable, and equipped with advanced sensors for situational awareness, ensuring that soldiers can maintain optimal performance under demanding conditions.

6.1.3 Benefits of Powered Exoskeletons

Enhanced Mobility and Independence

One of the primary benefits of powered exoskeletons is the significant improvement in mobility and independence for individuals with physical disabilities. These devices enable users to perform activities of daily living, such as walking, standing, and climbing stairs, with greater ease and confidence. By restoring mobility, exoskeletons contribute to improved quality of life, social participation, and overall well-being for individuals with mobility impairments.

Rehabilitation and Recovery

Powered exoskeletons offer a valuable tool for rehabilitation, providing repetitive and controlled movement patterns that are essential for neuroplasticity and functional recovery. The use of exoskeletons in rehabilitation programs accelerates the recovery process, enhances muscle strength, and promotes motor learning. Additionally, these devices offer real-time feedback and data collection, allowing therapists to monitor progress and adjust treatment plans accordingly.

Reduced Risk of Injuries

In industrial and occupational settings, powered exoskeletons reduce the physical strain on workers, minimizing the risk of musculoskeletal injuries and work-related accidents. By distributing the load and providing additional support, these devices alleviate stress on joints and muscles, enhancing worker safety and reducing absenteeism due to injuries. The ergonomic benefits of exoskeletons contribute to a healthier and more productive workforce.

6.1.4 Challenges and Considerations

Technical Limitations

Despite significant advancements, powered exoskeletons face technical challenges related to power supply, weight, and user comfort. Developing lightweight and durable power sources that provide extended battery life without compromising mobility remains a critical area of research. Additionally, ensuring that exoskeletons are comfortable for prolonged use, adaptable to different body types, and easy to don and doff is essential for user acceptance and widespread adoption.

Cost and Accessibility

The high cost of powered exoskeletons presents a barrier to accessibility for many individuals and organizations. The expense of developing, manufacturing, and maintaining these devices limits their availability, particularly in low-resource settings.

Efforts to reduce production costs, improve affordability, and secure funding for research and development are crucial to making powered exoskeletons more accessible to those in need.

Ethical and Social Implications

The deployment of powered exoskeletons raises ethical and social considerations, including issues related to equity, privacy, and the potential for dependency. Ensuring that these technologies are distributed equitably and do not exacerbate existing inequalities is paramount. Additionally, the collection and use of data generated by exoskeletons must adhere to stringent privacy and security standards to protect users' personal information. Addressing these ethical and social implications is essential for responsible and inclusive adoption of exoskeleton technologies.

6.1.5 Future Directions and Innovations

Integration with AI and Machine Learning

Future advancements in powered exoskeletons will likely involve deeper integration with AI and machine learning to enhance device adaptability, predictive capabilities, and user experience. AI algorithms can optimize movement patterns, predict user intentions, and provide personalized support based on individual needs and preferences. Machine learning models can continuously improve the performance of exoskeletons by analyzing user data and adapting to changes in mobility requirements.

Development of Soft Exoskeletons

Soft exoskeletons, also known as exosuits, represent a promising direction in exoskeleton technology. These devices use soft, flexible materials and actuators to provide assistance without the rigid structures of traditional exoskeletons. Soft exoskeletons offer greater comfort, ease of movement, and compatibility with various clothing and equipment. Research into soft robotics and wearable technologies will drive the development of lightweight, unobtrusive exoskeletons that enhance mobility with minimal impact on natural movement.

Expanded Applications and Accessibility

As technology advances and costs decrease, powered exoskeletons will find expanded applications across diverse fields. Beyond healthcare, industrial, and military uses, these devices may benefit areas such as sports, entertainment, and everyday mobility for aging populations. Increasing accessibility through affordable pricing, public funding, and inclusive design will ensure that the benefits of powered exoskeletons are available to a broader range of users, enhancing mobility and quality of life on a global scale.

6.2 SMART CLOTHING AND WEARABLE HEALTH MONITORS

Smart clothing and wearable health monitors represent a significant leap in the integration of technology with daily life, providing real-time health data, enhancing athletic performance, and promoting overall well-being.

Smart clothing and wearable health monitors represent a significant advancement in the intersection of technology and health, offering real-time data, enhanced athletic performance, and improved lifestyle management. While challenges related to data privacy, accuracy, and accessibility persist, ongoing innovations and ethical considerations will drive the future development of these technologies. The integration of AI, advancements in materials science, and the creation of personalized health ecosystems will further enhance the capabilities and impact of smart clothing and wearables, transforming the landscape of healthcare, fitness, and wellness in the years to come.

6.2.1 Development of Smart Clothing and Wearable Health Monitors

Early Innovations and Technological Evolution

The concept of integrating technology with clothing dates back to the early 20th century with the advent of wearable sensors for basic health monitoring. However, significant advancements in materials science, microelectronics, and wireless communication in the late 20th and early 21st centuries have propelled the development of sophisticated smart clothing and wearable health monitors. Early prototypes included simple pedometers and heart rate monitors, evolving into complex systems capable of tracking multiple physiological parameters.

Key Technologies

Modern smart clothing and wearable health monitors leverage several advanced technologies:

- **Sensors:** Embedded sensors detect and measure physiological signals such as heart rate, respiration, body temperature, and movement.
- **Microprocessors:** These process data from sensors and provide real-time feedback.
- **Wireless Communication:** Technologies like Bluetooth and Wi-Fi enable data transmission to smartphones, tablets, or cloud servers.
- **Advanced Materials:** Flexible and durable materials that integrate seamlessly with fabrics and withstand daily wear and tear.

6.2.2 Applications of Smart Clothing and Wearable Health Monitors

Healthcare Monitoring

Smart clothing and wearable health monitors are revolutionizing healthcare by enabling continuous, non-invasive monitoring of patients. These devices provide critical data on vital signs, allowing for early detection of medical conditions such as arrhythmias, sleep apnea, and chronic obstructive pulmonary disease (COPD). Wearable monitors can alert healthcare providers to abnormal readings, facilitating timely interventions and improving patient outcomes.

Fitness and Sports

In the fitness and sports industries, smart clothing and wearables enhance performance by providing athletes with detailed insights into their physical activity. Devices such as smartwatches, fitness bands, and smart shirts track metrics like heart rate, calories burned, steps taken, and sleep quality. Advanced wearables analyze biomechanical data, helping athletes optimize their training, prevent injuries, and achieve peak performance.

Lifestyle and Wellness

For the general population, smart clothing and wearables promote a healthier lifestyle by encouraging physical activity, monitoring sleep patterns, and managing stress. Devices can set daily activity goals, provide reminders to move, and offer relaxation techniques based on stress levels detected through physiological signals. This continuous feedback loop supports users in making healthier choices and improving their overall well-being.

6.2.3 Benefits of Smart Clothing and Wearable Health Monitors

Real-Time Health Data

One of the most significant benefits of smart clothing and wearable health monitors is the ability to collect and analyze real-time health data. Continuous monitoring provides a comprehensive picture of an individual's health, enabling proactive management of medical conditions and timely interventions. This real-time data can be shared with healthcare providers, allowing for personalized and informed medical care.

Enhanced Athletic Performance

Athletes and fitness enthusiasts benefit from the detailed performance data provided by wearables. By analyzing metrics such as heart rate variability, muscle activity, and movement patterns, athletes can tailor their training regimens to maximize efficiency and reduce the risk of injury. Smart clothing can also provide real-time feedback during workouts, helping users maintain optimal form and intensity.

Improved Lifestyle and Well-Being

Wearable health monitors support individuals in adopting healthier lifestyles by tracking daily activity, monitoring sleep quality, and managing stress. These devices offer personalized recommendations and reminders, helping users stay motivated and achieve their wellness goals. By promoting self-awareness and accountability, wearables contribute to long-term health and well-being.

6.2.4 Challenges and Considerations

Data Privacy and Security

The collection and transmission of sensitive health data raise significant privacy and security concerns. Ensuring that user data is protected from unauthorized access and breaches is paramount. Wearable device manufacturers must implement robust encryption, secure data storage, and strict access controls to safeguard personal information. Additionally, users should be informed about data usage policies and have control over their data sharing preferences.

Accuracy and Reliability

The accuracy and reliability of smart clothing and wearable health monitors are critical for their effectiveness. Variations in sensor quality, environmental factors, and user behavior can affect data accuracy. Continuous calibration, validation studies, and advancements in sensor technology are essential to improve the precision and dependability of these devices. Users and healthcare providers must be aware of the limitations and potential sources of error in wearable health data.

Cost and Accessibility

The cost of smart clothing and wearable health monitors can be a barrier to widespread adoption, particularly for individuals in low-income communities. Ensuring affordability and accessibility is crucial to maximizing the benefits of these technologies across diverse populations. Public health initiatives, insurance coverage, and partnerships with healthcare providers can help bridge the gap and make wearables more accessible to those who can benefit the most.

6.2.5 Future Directions and Innovations

Integration with Artificial Intelligence (AI)

The future of smart clothing and wearable health monitors lies in their integration with artificial intelligence (AI) and machine learning algorithms. AI can enhance the interpretation of health data, identify patterns, and provide predictive analytics for early disease detection and personalized health recommendations. AI-powered wearables can adapt to individual user behaviors and preferences, offering more accurate and relevant insights.

Development of Flexible and Biocompatible Materials

Advances in materials science will continue to drive innovation in smart clothing. The development of flexible, stretchable, and biocompatible materials will improve the comfort, durability, and functionality of wearable devices. These materials can be seamlessly integrated into everyday clothing, making health monitoring unobtrusive and more user-friendly.

Expanded Applications in Healthcare

Future wearable health monitors will likely play a more prominent role in clinical settings, supporting remote patient monitoring, chronic disease management, and post-operative care. Integration with electronic health records (EHRs) and telemedicine platforms will enhance the continuity of care, enabling healthcare providers to monitor patients in real-time and adjust treatment plans based on up-to-date health data.

Personalized Health and Wellness Ecosystems

The development of comprehensive health and wellness ecosystems that integrate wearables, mobile apps, and cloud-based platforms will provide users with a holistic view of their health. These ecosystems will offer personalized insights, track progress over time, and connect users with healthcare professionals and wellness resources. By fostering a collaborative approach to health management, these ecosystems will empower individuals to take proactive control of their well-being.

6.3 HUMAN AUGMENTATION FOR OCCUPATIONAL TASKS

Human augmentation involves using technology to enhance human capabilities, making it possible to perform occupational tasks more efficiently, safely, and effectively. From wearable exoskeletons and augmented reality (AR) glasses to advanced brain-machine interfaces, human augmentation is revolutionizing various industries by augmenting human strength, dexterity, cognition, and sensory perception.

Human augmentation for occupational tasks represents a transformative advancement in enhancing human capabilities, improving productivity, and ensuring worker safety. While there are challenges related to cost, ethical concerns, and technological limitations, ongoing research, innovation, and collaboration will drive the development and adoption of these technologies. As we look to the future, the integration of advanced AI, the development of more comfortable wearables, and the establishment of robust policy frameworks will be critical in realizing the full potential of human augmentation in the workplace.

6.3.1 Types of Human Augmentation Technologies

Wearable Exoskeletons

Wearable exoskeletons are external frameworks that support and enhance human movement. These devices can be passive, relying on mechanical structures, or active, incorporating powered actuators to amplify human strength and endurance. They are commonly used in industries such as manufacturing, construction, and healthcare to reduce physical strain and prevent injuries.

Augmented Reality (AR) and Virtual Reality (VR)

AR and VR technologies provide enhanced visual information to workers through devices such as smart glasses or headsets. AR overlays digital information onto the real world, aiding tasks like assembly, maintenance, and training by providing real-time instructions and guidance. VR creates immersive simulations for training, allowing workers to practice complex tasks in a controlled, risk-free environment.

Brain-Machine Interfaces (BMIs)

BMIs enable direct communication between the brain and external devices, allowing for control of machinery or computers through neural signals. This technology can assist individuals with disabilities in performing occupational tasks and has potential applications in precision tasks that require fine motor skills and rapid response times.

Haptic Feedback Devices

Haptic feedback devices provide tactile sensations to the user, enhancing the sense of touch and enabling precise control of tools and machinery. These devices are valuable in fields such as surgery, where surgeons can feel tissue resistance through robotic instruments, and in remote operation of equipment in hazardous environments.

6.3.2 Applications in Different Occupational Settings

Manufacturing and Assembly

In manufacturing and assembly lines, wearable exoskeletons reduce physical fatigue and injury risks by supporting the body during repetitive and strenuous tasks. AR glasses provide workers with real-time visual instructions and quality control alerts, increasing accuracy and efficiency. Haptic feedback devices enhance the precision of assembly tasks, especially in electronics and small-scale manufacturing.

Construction and Heavy Industry

Exoskeletons and powered suits assist construction workers in lifting heavy materials and maintaining ergonomic postures, thereby reducing the risk of musculoskeletal disorders. AR technologies can display construction plans directly in the worker's field of view, facilitating accurate measurements and alignment. VR simulations provide training for operating heavy machinery, ensuring that workers are well-prepared before handling real equipment.

Healthcare and Rehabilitation

In healthcare, exoskeletons help nurses and caregivers lift and transfer patients, minimizing the risk of back injuries. Surgeons use AR and VR for preoperative planning and intraoperative guidance, enhancing surgical precision and outcomes. BMIs and haptic feedback devices enable more effective rehabilitation for patients with neurological impairments by providing interactive and responsive therapy sessions.

Logistics and Warehousing

Logistics and warehousing operations benefit from exoskeletons that support workers in lifting and moving heavy packages, reducing the incidence of workplace injuries. AR glasses assist with inventory management by displaying real-time data on stock levels and locations, streamlining the picking and packing processes. VR training modules prepare workers for operating forklifts and other machinery safely and efficiently.

Military and Defense

In military applications, exoskeletons enhance soldiers' physical capabilities, allowing them to carry heavier loads and endure longer missions. AR systems provide battlefield situational awareness, displaying critical information such as enemy positions and navigation routes. BMIs have potential applications in controlling drones and other unmanned systems, enabling rapid and precise responses to dynamic combat situations.

6.3.3 Benefits of Human Augmentation in Occupational Tasks

Enhanced Productivity and Efficiency

Human augmentation technologies improve productivity and efficiency by enabling workers to perform tasks more quickly and accurately. Real-time guidance through AR and VR reduces errors and accelerates learning curves, while exoskeletons and BMIs allow for the execution of physically demanding and precision tasks with greater ease.

Improved Worker Safety and Health

By reducing physical strain and fatigue, wearable exoskeletons help prevent work-related injuries and musculoskeletal disorders. AR and VR provide safer training environments, minimizing risks associated with on-the-job learning. Haptic feedback devices enhance control in delicate or hazardous tasks, ensuring that workers can operate safely and confidently.

Increased Accessibility and Inclusion

Human augmentation technologies make it possible for individuals with disabilities to participate more fully in the workforce. BMIs and exoskeletons enable those with mobility impairments to perform tasks that would otherwise be challenging or impossible. This promotes greater inclusion and diversity in various occupational settings.

6.3.4 Challenges and Considerations

Cost and Accessibility

The high cost of human augmentation technologies can be a barrier to widespread adoption, particularly for small and medium-sized enterprises. Ensuring that these technologies are affordable and accessible is crucial for maximizing their benefits. This may require public funding, subsidies, and partnerships between technology providers and industry stakeholders.

Ethical and Privacy Concerns

The use of human augmentation technologies raises ethical and privacy concerns, particularly regarding data collection and surveillance. Employers must ensure that the data collected from wearable devices is used responsibly and that workers' privacy is protected. Transparent policies and regulations are needed to address these concerns and build trust among employees.

Technological Limitations

Current human augmentation technologies have limitations related to battery life, durability, and user comfort. Continuous research and development are needed to address these issues and enhance the reliability and usability of these devices. Ensuring that technologies are adaptable to different work environments and individual needs is also essential for their effectiveness.

Training and Adaptation

Integrating human augmentation technologies into the workplace requires adequate training for workers to use these devices effectively and safely. Organizations must invest in comprehensive training programs and provide ongoing support to ensure smooth adaptation. Additionally, fostering a positive attitude towards technology adoption and addressing potential resistance from workers are crucial for successful implementation.

6.3.5 Future Prospects and Innovations

Advanced AI Integration

The integration of artificial intelligence (AI) with human augmentation technologies will enhance their functionality and adaptability. AI algorithms can provide personalized assistance, predict potential hazards, and optimize task performance based on real-time data. This will further improve efficiency, safety, and user experience in various occupational settings.

Development of Soft and Flexible Wearables

Future innovations in materials science will lead to the development of soft and flexible wearable devices that are more comfortable and less obtrusive. These advancements will improve user compliance and expand the range of tasks that can be augmented. Soft exoskeletons and wearable sensors embedded in clothing are examples of this trend.

Broader Applications and Collaboration

Human augmentation technologies will find broader applications across diverse fields such as agriculture, education, and emergency response. Collaboration between technology developers, industry leaders, and regulatory bodies will be essential to drive innovation and ensure that these technologies are implemented ethically and effectively.

Policy and Regulatory Frameworks

As human augmentation technologies become more prevalent, comprehensive policy and regulatory frameworks will be needed to address ethical, safety, and privacy concerns. These frameworks should promote responsible innovation, protect workers' rights, and ensure that the benefits of human augmentation are distributed equitably across society.

CHAPTER 7
COLLABORATIVE ROBOTS AND COBOTS

7.1 THE RISE OF COLLABORATIVE ROBOTICS

The landscape of robotics has evolved dramatically over the past few decades, moving from isolated, industrial machines to highly interactive, human-friendly systems known as collaborative robots, or cobots. These advanced robots are designed to work alongside humans in various settings, enhancing productivity, safety, and efficiency.

The rise of collaborative robotics marks a significant milestone in the evolution of automation, offering transformative benefits across various industries. Cobots enhance productivity, safety, and efficiency by working alongside humans, leveraging advanced technologies such as AI, sensors, and human-robot interfaces. While challenges related to integration, cost, and workforce adaptation persist, ongoing innovations and a collaborative approach will drive the future of cobots, unlocking new possibilities and reshaping the world of work.

7.1.1 Development of Collaborative Robotics

Early Robotics and Industrial Automation

The initial wave of industrial robots, beginning in the mid-20th century, focused on automating repetitive and hazardous tasks in manufacturing environments. These robots operated in isolation from human workers due to safety concerns and the lack of sophisticated sensors and control systems. Early industrial robots were primarily used for tasks such as welding, painting, and assembly, significantly improving production efficiency but offering limited flexibility.

Emergence of Collaborative Robotics

The concept of collaborative robotics emerged as advancements in sensors, artificial intelligence (AI), and machine learning enabled robots to safely and effectively interact with humans. Unlike traditional industrial robots, cobots are designed with safety features such as force and torque sensors, allowing them to detect and respond to human presence. The development of lightweight, flexible, and user-friendly robots marked a significant shift towards more versatile and adaptable automation solutions.

Technological Foundations

Several key technologies underpin the rise of collaborative robotics:

- **Advanced Sensors:** Enable cobots to perceive their environment, detect obstacles, and interact safely with humans.
- **AI and Machine Learning:** Enhance the adaptability and decision-making capabilities of cobots, allowing them to perform complex tasks and learn from experience.
- **Human-Robot Interfaces (HRIs):** Facilitate intuitive and seamless interaction between humans and robots through touchscreens, voice commands, and gesture recognition.

7.1.2 Applications of Collaborative Robotics

Manufacturing and Assembly

In manufacturing, cobots are transforming assembly lines by working alongside human workers to perform tasks such as picking, placing, and assembling components. Cobots enhance efficiency and precision, reduce ergonomic strain on workers, and enable flexible production lines that can be quickly reconfigured for different products. Companies like Universal Robots and KUKA have pioneered the use of cobots in automotive, electronics, and consumer goods manufacturing.

Healthcare and Medical Applications

Cobots are increasingly being utilized in healthcare settings for tasks ranging from surgery to patient care. Surgical robots, such as the da Vinci Surgical System, provide enhanced precision and control, enabling minimally invasive procedures with improved outcomes. In rehabilitation, cobots assist patients with physical therapy exercises, providing consistent and adjustable support to aid recovery. Additionally, cobots are used in laboratories for tasks such as sample handling and analysis, increasing throughput and accuracy.

Logistics and Warehousing

In logistics and warehousing, cobots streamline operations by automating repetitive tasks such as sorting, packing, and transporting goods. Autonomous mobile robots (AMRs) navigate warehouse environments, collaborating with human workers to fulfill orders quickly and accurately. Companies like Amazon and DHL have integrated cobots into their supply chain operations, reducing costs and improving efficiency.

Retail and Customer Service

Cobots are also making inroads into the retail sector, where they assist with inventory management, shelf restocking, and customer service. Retail robots, such as SoftBank's Pepper, interact with customers to provide information, answer queries, and enhance the shopping experience. In addition, cobots help manage inventory by performing regular stock checks and ensuring products are correctly placed and available.

Agriculture and Food Industry

In agriculture, cobots are deployed for tasks such as planting, harvesting, and sorting crops. These robots can operate around the clock, improving yield and reducing labor costs.

In the food industry, cobots assist with food preparation, packaging, and quality control, ensuring consistent quality and hygiene standards. The use of cobots in these sectors addresses labor shortages and increases productivity.

7.1.3 Benefits of Collaborative Robotics

Increased Productivity and Efficiency

Cobots enhance productivity and efficiency by performing repetitive, precise, and time-consuming tasks, allowing human workers to focus on more complex and value-added activities.

This collaboration between humans and robots optimizes workflow, reduces downtime, and increases overall output.

Improved Safety and Ergonomics

Cobots are designed with built-in safety features that minimize the risk of accidents and injuries. By taking over physically demanding and hazardous tasks, cobots reduce the ergonomic strain on workers, promoting a safer and healthier work environment. This is particularly beneficial in industries with high rates of musculoskeletal disorders.

Flexibility and Adaptability

Unlike traditional industrial robots, cobots are highly flexible and adaptable to different tasks and environments. They can be easily programmed and reprogrammed, allowing businesses to quickly respond to changing demands and production requirements. This adaptability makes cobots suitable for small and medium-sized enterprises (SMEs) that require scalable automation solutions.

Enhanced Quality and Consistency

Cobots ensure high levels of precision and consistency in task execution, leading to improved product quality and reduced error rates. In industries such as electronics and pharmaceuticals, where accuracy is critical, cobots play a vital role in maintaining stringent quality standards.

7.1.4 Challenges and Considerations

Integration and Compatibility

Integrating cobots into existing workflows and systems can be challenging, particularly in legacy environments with outdated infrastructure. Ensuring compatibility with existing equipment, software, and processes requires careful planning and investment in integration solutions.

Initial Costs and ROI

While cobots offer long-term cost savings, the initial investment can be significant, especially for SMEs. Assessing the return on investment (ROI) and justifying the expenditure requires a thorough understanding of the potential benefits and cost savings associated with cobot deployment.

Workforce Training and Adaptation

Successful implementation of cobots depends on adequately training the workforce to interact with and operate these advanced systems.

Workers may need to develop new skills and adapt to changing roles, which can be met with resistance. Providing comprehensive training and support is essential for a smooth transition.

Ethical and Social Implications

The rise of collaborative robotics raises ethical and social considerations, including concerns about job displacement and the impact on the workforce. While cobots can enhance productivity and create new job opportunities, there is a need to address potential job losses and ensure that workers are equipped with the skills required for emerging roles.

7.1.5 Future Prospects and Innovations

Advancements in AI and Machine Learning

Future developments in AI and machine learning will further enhance the capabilities of cobots, enabling them to perform more complex tasks, learn from experience, and adapt to new environments. These advancements will lead to more autonomous and intelligent collaborative robots that can seamlessly integrate into various industries.

Enhanced Human-Robot Interaction

Innovations in human-robot interaction (HRI) will improve the way humans and cobots communicate and collaborate. Advances in natural language processing, gesture recognition, and tactile feedback will make interactions more intuitive and efficient, fostering greater acceptance and usability of cobots in the workplace.

Expansion into New Industries

As technology continues to evolve, cobots will find applications in a broader range of industries, including education, entertainment, and environmental monitoring. The versatility and adaptability of cobots will drive their adoption in diverse fields, creating new opportunities for innovation and growth.

Ethical Frameworks and Workforce Development

Addressing the ethical and social implications of collaborative robotics will be crucial for their sustainable development. Policymakers, industry leaders, and educational institutions must collaborate to develop ethical frameworks, promote responsible innovation, and invest in workforce development programs that prepare individuals for the future of work.

7.2 HUMAN-COBOT INTERACTION IN MANUFACTURING

The integration of collaborative robots, or cobots, into manufacturing processes represents a significant advancement in industrial automation. Unlike traditional industrial robots that operate in isolation, cobots are designed to work alongside human workers, enhancing productivity, safety, and flexibility.

The integration of collaborative robots into manufacturing processes represents a significant advancement in industrial automation. Cobots enhance productivity, safety, and flexibility by working alongside human workers, leveraging advanced technologies such as AI, sensors, and human-robot interfaces. While challenges related to integration, cost, and workforce adaptation persist, ongoing innovations and a collaborative approach will drive the future of human-cobot interaction. As cobots continue to evolve, they will play a pivotal role in shaping the future of manufacturing and beyond, unlocking new possibilities and transforming the world of work.

7.2.1 Evolution of Collaborative Robots

From Industrial Robots to Cobots

The history of industrial robots dates back to the 1960s when they were introduced to automate repetitive and dangerous tasks in manufacturing. These robots operated within safety cages, separated from human workers due to the risk of injury. Over time, advancements in sensors, artificial intelligence (AI), and machine learning enabled the development of cobots, which are designed to safely collaborate with humans in shared workspaces.

Technological Innovations

Cobots are equipped with advanced technologies that facilitate safe and efficient human-robot interaction:

- **Sensors and Vision Systems:** These allow cobots to detect and respond to human presence, avoiding collisions and ensuring safe operation.
- **AI and Machine Learning:** Enhance the cobots' ability to learn from human behavior, adapt to new tasks, and improve performance over time.
- **Human-Robot Interfaces (HRIs):** Facilitate intuitive interaction through touchscreens, voice commands, and gesture recognition.

7.2.2 Applications in Manufacturing

Assembly and Production Lines

Cobots are widely used in assembly and production lines to perform tasks such as picking, placing, and assembling components. They work alongside human workers, enhancing efficiency and precision. For example, cobots can handle repetitive tasks while humans focus on more complex and decision-intensive activities, leading to a more balanced and productive workflow.

Quality Control and Inspection

Cobots play a crucial role in quality control and inspection processes. Equipped with high-resolution cameras and sensors, they can detect defects and inconsistencies with high accuracy. This capability ensures that products meet stringent quality standards and reduces the likelihood of defective items reaching the market.

Material Handling and Logistics

In manufacturing environments, cobots assist with material handling tasks, including sorting, packaging, and transporting goods. Their ability to work continuously without fatigue improves operational efficiency and reduces the risk of workplace injuries associated with manual handling of heavy or hazardous materials.

Machine Tending

Cobots are employed in machine tending operations, where they load and unload parts from machinery such as CNC machines and injection molding equipment. This automation reduces downtime, increases production throughput, and allows human workers to focus on tasks that require more skill and judgment.

7.2.3 Benefits of Human-Cobot Interaction

Enhanced Productivity

The collaboration between humans and cobots leads to significant productivity gains. Cobots can operate around the clock, performing repetitive tasks with high precision and consistency. This allows human workers to concentrate on tasks that require creativity, problem-solving, and critical thinking, thereby optimizing overall workflow.

Improved Safety and Ergonomics

Cobots are designed with safety features such as force and torque sensors that enable them to detect and react to human presence. This ensures that they can work safely alongside humans without the need for physical barriers. By taking over physically demanding and repetitive tasks, cobots reduce the risk of musculoskeletal injuries and improve workplace ergonomics.

Flexibility and Adaptability

Cobots are highly flexible and can be easily programmed and reprogrammed to perform different tasks. This adaptability makes them ideal for small and medium-sized enterprises (SMEs) that require scalable automation solutions. Cobots can quickly switch between tasks and production lines, accommodating varying production requirements and reducing downtime.

Quality and Precision

Cobots enhance the quality and precision of manufacturing processes. Their ability to perform tasks with consistent accuracy reduces the likelihood of errors and defects. This is particularly beneficial in industries such as electronics and pharmaceuticals, where precision is critical to product quality and safety.

7.2.4 Challenges in Human-Cobot Interaction

Integration and Compatibility

Integrating cobots into existing manufacturing processes can be challenging, particularly in facilities with legacy equipment and infrastructure. Ensuring compatibility between cobots and existing systems requires careful planning and investment in integration technologies. Manufacturers must also consider the potential impact on workflow and production schedules during the transition period.

Initial Costs and ROI

The initial investment in cobots can be significant, especially for smaller companies. Assessing the return on investment (ROI) involves considering not only the purchase cost but also the expenses related to installation, training, and maintenance. However, the long-term benefits of increased productivity and reduced labor costs can justify the initial expenditure.

Workforce Training and Adaptation

Successful implementation of cobots requires comprehensive training for human workers. Employees need to learn how to interact with, program, and troubleshoot cobots. Additionally, there may be resistance to change from workers who fear job displacement or are unfamiliar with new technologies. Addressing these concerns through effective communication and training programs is essential for smooth adaptation.

Ethical and Social Considerations

The rise of cobots raises ethical and social questions, particularly regarding job displacement and the future of work. While cobots can enhance productivity and create new job opportunities, there is a need to address potential job losses and ensure that workers are equipped with the skills required for emerging roles. Developing policies and strategies to support workforce transition is crucial.

7.2.5 Future Directions in Human-Cobot Interaction

Advanced AI and Machine Learning

The future of cobots lies in the continued integration of advanced AI and machine learning algorithms. These technologies will enable cobots to perform more complex tasks, learn from human behavior, and improve their performance over time. AI-driven cobots will be able to make autonomous decisions, adapt to dynamic environments, and collaborate more effectively with human workers.

Enhanced Human-Robot Interfaces

Improvements in human-robot interfaces will facilitate more intuitive and seamless interaction between humans and cobots. Advances in natural language processing, gesture recognition, and haptic feedback will enable workers to communicate and collaborate with cobots more efficiently. This will enhance the overall user experience and increase acceptance of cobots in the workplace.

Expanded Applications and Industries

As technology continues to evolve, cobots will find applications in a broader range of industries beyond manufacturing, including healthcare, agriculture, and services. Their versatility and adaptability will drive adoption in diverse fields, creating new opportunities for innovation and growth. For example, in healthcare, cobots could assist with patient care and medical procedures, while in agriculture, they could help with planting and harvesting crops.

Ethical Frameworks and Workforce Development

Addressing the ethical and social implications of cobot integration is essential for sustainable development. Policymakers, industry leaders, and educational institutions must collaborate to develop ethical frameworks, promote responsible innovation, and invest in workforce development programs. These initiatives will ensure that the benefits of cobot technology are shared equitably and that workers are prepared for the future of work.

7.3 COBOTS IN HEALTHCARE AND SERVICE INDUSTRIES

The advent of collaborative robots, or cobots, has brought significant advancements across various sectors, including healthcare and service industries. Unlike traditional robots that operate in isolation, cobots are designed to work alongside humans, enhancing efficiency, safety, and the quality of services provided.

The integration of collaborative robots into healthcare and service industries marks a significant advancement in automation and human-robot collaboration. Cobots enhance efficiency, safety, and service quality by performing tasks with precision and consistency.

While challenges related to integration, training, and ethical concerns persist, ongoing innovations and a collaborative approach will drive the future of cobots. As these technologies continue to evolve, they will play a pivotal role in transforming healthcare and service delivery, unlocking new possibilities, and improving the overall quality of life.

7.3.1 Introduction to Cobots

Definition and Characteristics

Collaborative robots, commonly known as cobots, are robots designed to work in close collaboration with human workers. They are equipped with advanced sensors, artificial intelligence (AI), and machine learning capabilities, allowing them to perform tasks with precision and safety. Unlike traditional industrial robots, cobots are typically smaller, more flexible, and easier to program, making them suitable for various environments, including healthcare and service industries.

Evolution and Adoption

The concept of cobots emerged from the need to create robots that can safely and efficiently interact with humans in shared workspaces. Over the past decade, significant technological advancements have accelerated their adoption. Today, cobots are employed in diverse applications, ranging from manufacturing and logistics to healthcare and customer service.

7.3.2 Applications of Cobots in Healthcare

Surgical Assistance

One of the most notable applications of cobots in healthcare is in surgical assistance. Cobots like the da Vinci Surgical System provide surgeons with enhanced precision, control, and dexterity during minimally invasive procedures. These cobots translate the surgeon's hand movements into smaller, more precise movements of tiny instruments inside the patient's body, reducing the risk of human error and improving surgical outcomes.

Rehabilitation and Therapy

Cobots are also playing a crucial role in rehabilitation and physical therapy. Robots like the Lokomat and ArmeoSpring are designed to assist patients with movement impairments, providing consistent and adjustable support during therapy sessions.

These cobots help patients regain mobility and strength through repetitive, guided exercises, leading to faster and more effective rehabilitation.

Patient Care and Assistance

In hospitals and care facilities, cobots are being used to assist with routine tasks such as lifting and transferring patients, delivering medications, and monitoring vital signs.

Robots like the TUG and Moxi can autonomously navigate hospital environments, reducing the workload on healthcare staff and allowing them to focus on more critical patient care activities. These cobots help improve efficiency and ensure timely delivery of care services.

Laboratory Automation

Cobots are also transforming laboratory operations by automating repetitive and time-consuming tasks such as sample handling, pipetting, and analysis. Robots like the Hamilton STAR and the Cobas 6800 can process large volumes of samples with high accuracy and speed, reducing the risk of contamination and human error. This automation frees up laboratory technicians to focus on more complex tasks and research activities.

7.3.3 Applications of Cobots in Service Industries

Customer Service and Hospitality

Cobots are increasingly being employed in customer service and hospitality sectors to enhance the customer experience. Robots like SoftBank's Pepper and Savioke's Relay can greet guests, provide information, and deliver items to rooms. These cobots can interact with customers using natural language processing, making them suitable for roles in hotels, restaurants, and retail environments. Their presence helps improve service efficiency and customer satisfaction.

Retail and Inventory Management

In retail, cobots assist with inventory management, shelf restocking, and customer assistance. Robots like Tally and Bossa Nova can autonomously scan shelves to monitor stock levels, identify misplaced items, and generate restocking alerts. This real-time data helps retailers maintain accurate inventory, reduce out-of-stock situations, and ensure that products are correctly displayed for customers.

Cleaning and Maintenance

Cobots are also being used for cleaning and maintenance tasks in various service environments. Robots like the Whiz floor-cleaning robot and the Xenex Germ-Zapping Robot can autonomously clean and disinfect surfaces, reducing the workload on human staff and ensuring a high standard of hygiene. These cobots are particularly valuable in settings like hospitals, airports, and shopping malls, where cleanliness is crucial.

Delivery and Logistics

In the logistics sector, cobots are employed to optimize delivery and supply chain operations. Autonomous mobile robots (AMRs) like those developed by Starship Technologies and Nuro can transport goods within facilities and deliver packages to customers' doorsteps. These cobots enhance the efficiency and accuracy of delivery services, reducing operational costs and improving customer satisfaction.

7.3.4 Benefits of Cobots in Healthcare and Service Industries

Increased Efficiency and Productivity

Cobots significantly enhance efficiency and productivity by automating repetitive and time-consuming tasks. In healthcare, this allows medical professionals to focus on patient care and complex procedures, while in service industries, it leads to faster and more accurate service delivery. The consistent performance of cobots ensures high-quality outcomes and reduces the likelihood of human error.

Improved Safety and Ergonomics

Cobots are designed with safety features such as force sensors and collision detection, allowing them to operate safely alongside humans. In healthcare, this reduces the risk of injuries to both patients and staff during tasks such as patient lifting and transferring. In service industries, cobots take over physically demanding tasks, improving workplace ergonomics and reducing the risk of musculoskeletal disorders.

Enhanced Service Quality

The precision and consistency of cobots enhance the quality of services provided. In healthcare, cobots improve surgical outcomes and rehabilitation effectiveness. In service industries, they ensure accurate inventory management, timely delivery of goods, and consistent cleaning standards. This leads to higher customer satisfaction and better overall service experiences.

Cost Savings and ROI

While the initial investment in cobots can be significant, the long-term cost savings and return on investment (ROI) are substantial. Cobots reduce labor costs, minimize errors, and increase operational efficiency, leading to significant financial benefits over time. Additionally, the scalability and flexibility of cobots allow businesses to adapt to changing demands without extensive reconfiguration costs.

7.3.5 Challenges and Considerations

Integration and Compatibility

Integrating cobots into existing workflows and systems can be challenging. Ensuring compatibility with current infrastructure and processes requires careful planning and investment.

In healthcare, this may involve integrating cobots with electronic health records (EHR) systems and other medical devices. In service industries, it may require coordination with inventory management and customer relationship management (CRM) systems.

Workforce Training and Adaptation

Successful implementation of cobots requires comprehensive training for human workers. Employees need to understand how to operate, program, and troubleshoot cobots. Additionally, there may be resistance to change from workers who fear job displacement or are unfamiliar with new technologies. Providing effective training and addressing these concerns through transparent communication is essential for smooth adaptation.

Ethical and Privacy Concerns

The use of cobots in healthcare and service industries raises ethical and privacy concerns, particularly regarding data collection and patient/customer interactions. Ensuring that data is collected and used responsibly, and that privacy is protected, is crucial. Developing clear ethical guidelines and regulatory frameworks will help address these concerns and build trust among stakeholders.

7.3.6 Future Prospects and Innovations

Advanced AI and Machine Learning

Future advancements in AI and machine learning will further enhance the capabilities of cobots, enabling them to perform more complex tasks, learn from experience, and adapt to dynamic environments. This will lead to more intelligent and autonomous cobots that can provide even greater benefits in healthcare and service industries.

Enhanced Human-Robot Interaction

Improvements in human-robot interaction (HRI) technologies will facilitate more intuitive and seamless collaboration between humans and cobots. Advances in natural language processing, gesture recognition, and tactile feedback will make interactions more efficient and user-friendly. This will enhance the overall user experience and increase acceptance of cobots in various settings.

Broader Applications and Adoption

As technology continues to evolve, cobots will find applications in a wider range of industries and tasks. In healthcare, they may assist with telemedicine, remote diagnostics, and personalized care. In service industries, cobots could support more complex customer service interactions and personalized shopping experiences. The versatility and adaptability of cobots will drive their adoption across diverse fields, creating new opportunities for innovation and growth.

CHAPTER 8

AUGMENTING CREATIVITY AND INNOVATION

8.1 HUMAN-AI COLLABORATION IN CREATIVE INDUSTRIES

The advent of artificial intelligence (AI) has significantly impacted various sectors, including the creative industries. AI has evolved from performing routine tasks to participating in complex, creative processes alongside human professionals. This collaboration between human creativity and AI's computational power opens up new possibilities in art, music, design, literature, and more.

The collaboration between human creativity and artificial intelligence is transforming the creative industries, offering new possibilities and opportunities. AI enhances creativity, increases efficiency, and provides data-driven insights, making it a valuable tool for artists, musicians, writers, designers, and other creative professionals. However, challenges related to authenticity, job displacement, ethical use, and intellectual property rights must be addressed to ensure the responsible and equitable integration of AI in creative processes. As technology continues to advance, the future of human-AI collaboration in creative industries promises to be exciting and transformative, unlocking new dimensions of creativity and innovation.

8.1.1 Introduction to AI in Creative Industries

Definition and Scope of AI

Artificial intelligence refers to the simulation of human intelligence in machines that are programmed to think, learn, and adapt. In the context of creative industries, AI involves using algorithms and computational power to generate, enhance, or assist in creating artistic content. This can include anything from generating music and visual art to designing products and writing literature.

Historical Context and Evolution

The use of AI in creative processes is not entirely new. Early experiments with computer-generated art and music date back to the mid-20th century. However, recent advancements in machine learning, particularly deep learning and neural networks, have significantly expanded AI's capabilities. Today, AI can analyze vast amounts of data, recognize patterns, and generate content that rivals human creativity.

8.1.2 Applications of AI in Creative Industries

Visual Arts

AI has made significant strides in visual arts through techniques like style transfer, where the style of one artwork is applied to another image. Tools like DeepArt and Google's DeepDream have gained popularity for their ability to create visually stunning pieces. AI is also used for generating original artworks. For instance, the AI model GAN (Generative Adversarial Network) has been employed to create novel images that are exhibited in galleries and sold at auctions.

Music Composition

In music, AI assists in composing, arranging, and producing music. AI algorithms analyze existing musical works to understand patterns and styles, enabling them to compose new pieces in a similar vein. Applications like AIVA (Artificial Intelligence Virtual Artist) and OpenAI's MuseNet can create music in various genres, from classical to jazz and pop. Musicians use these tools to spark inspiration, develop ideas, and even produce entire albums.

Literature and Writing

AI's role in literature includes generating text, assisting with plot development, and even writing poetry. Tools like OpenAI's GPT-3 can generate coherent and contextually relevant text based on prompts. Authors and writers use AI to brainstorm ideas, develop characters, and overcome writer's block. AI-generated poetry and short stories are also gaining recognition for their creativity and depth.

Design and Fashion

In design and fashion, AI helps in creating innovative designs and predicting trends. AI algorithms analyze past and current fashion trends to forecast future styles, assisting designers in creating collections that resonate with consumers. AI tools like GANs generate new fashion designs by combining elements of existing styles. In product design, AI aids in optimizing designs for functionality and aesthetics, using generative design techniques.

Film and Animation

AI is revolutionizing film and animation by automating time-consuming tasks and enhancing creative processes. AI-powered tools assist in scriptwriting, storyboarding, and editing. They can also generate realistic visual effects and animations. For example, AI algorithms help in creating lifelike CGI characters and scenes, reducing the workload on animators and speeding up production timelines.

8.1.3 Benefits of Human-AI Collaboration

Enhanced Creativity

The collaboration between humans and AI leads to enhanced creativity by combining human intuition and emotional depth with AI's analytical and generative capabilities. AI serves as a powerful tool for brainstorming, idea generation, and experimenting with new concepts. This synergy allows creators to push the boundaries of their craft and explore uncharted creative territories.

Increased Efficiency

AI automates repetitive and time-consuming tasks, allowing creative professionals to focus on more complex and value-added activities. In fields like animation and design, AI tools accelerate the creative process, enabling faster production cycles and reducing the time from concept to final product. This increased efficiency helps artists and designers meet tight deadlines and manage larger workloads.

Data-Driven Insights

AI provides data-driven insights that inform creative decisions. By analyzing vast amounts of data on consumer preferences, trends, and historical patterns, AI helps creators understand their audience better and tailor their work accordingly. This data-driven approach enhances the relevance and appeal of creative outputs, increasing their chances of success in the market.

Democratization of Creativity

AI-powered tools democratize creativity by making advanced creative capabilities accessible to a broader audience. Aspiring artists, musicians, and writers can leverage AI tools to create professional-quality work without extensive training or resources. This democratization fosters a more inclusive creative landscape, where diverse voices and perspectives can thrive.

8.1.4 Challenges and Ethical Considerations

Authenticity and Originality

One of the primary challenges in human-AI collaboration is ensuring authenticity and originality. Critics argue that AI-generated content may lack the emotional depth and unique perspective that characterize human creativity. The question of authorship and originality arises when AI significantly contributes to a creative work, challenging traditional notions of creativity and intellectual property.

Job Displacement

The rise of AI in creative industries raises concerns about job displacement. As AI tools become more capable, there is a fear that they might replace human professionals in certain roles. However, many experts believe that AI will augment rather than replace human creativity, creating new roles and opportunities that require a combination of human and AI skills.

Ethical Use of AI

Ethical considerations surrounding the use of AI in creative industries include issues of bias, transparency, and accountability. AI algorithms can inadvertently perpetuate biases present in the data they are trained on, leading to biased outputs. Ensuring transparency in how AI-generated content is created and used is crucial for maintaining trust. Additionally, accountability for the ethical use of AI in creative processes lies with both developers and users of AI tools.

Intellectual Property Rights

The question of intellectual property rights becomes complex when AI is involved in the creative process. Determining who owns the rights to AI-generated content and how it can be used and monetized requires clear legal frameworks. Current intellectual property laws may need to be adapted to address these new challenges and ensure fair recognition and compensation for human creators and AI developers.

8.1.5 Future Prospects and Innovations

Advancements in AI Technology

Future advancements in AI technology will further enhance its capabilities in creative industries. Improvements in natural language processing, computer vision, and machine learning will enable AI to generate even more sophisticated and nuanced creative content. These advancements will push the boundaries of what is possible in art, music, literature, and design, opening up new avenues for exploration and innovation.

Human-AI Co-Creation

The future of human-AI collaboration lies in co-creation, where humans and AI work together seamlessly to create innovative and impactful work. This collaborative approach leverages the strengths of both human creativity and AI's computational power. Co-creation will lead to the development of new creative forms and genres that blend human intuition with AI's analytical capabilities.

Integration across Industries

AI's role in creative industries will continue to expand as it integrates with other technologies such as virtual reality (VR), augmented reality (AR), and blockchain. These integrations will create immersive and interactive creative experiences, revolutionizing how art, music, and entertainment are produced and consumed.

Blockchain technology can also address intellectual property and copyright issues by providing transparent and secure ownership records for AI-generated content.

Education and Training

As AI becomes more prevalent in creative industries, education and training programs will need to evolve to prepare future creators for this new landscape. Teaching students how to use AI tools effectively, understand their capabilities, and navigate ethical considerations will be essential. Educational institutions will play a crucial role in equipping the next generation of artists, musicians, writers, and designers with the skills needed to thrive in a human-AI collaborative environment.

8.2 AUGMENTED REALITY FOR DESIGN AND CREATIVITY

Augmented Reality (AR) is revolutionizing the way designers and creatives approach their work. By overlaying digital content onto the physical world, AR enables a more interactive, immersive, and engaging experience. Augmented Reality is transforming the design and creativity landscape by providing enhanced visualization, increased engagement, and improved collaboration. While there are challenges to address, such as technical limitations and privacy concerns, the benefits of AR are significant. As technology continues to advance, AR will become an even more integral part of the creative process, opening up new possibilities and expanding the boundaries of what is possible in design and art. The future of AR in creative industries is bright, promising to revolutionize how we create, experience, and interact with the world around us.

8.2.1 Introduction to Augmented Reality

Definition and Mechanism

Augmented Reality (AR) is a technology that enhances the real-world environment by overlaying digital information onto it. Unlike Virtual Reality (VR), which creates an entirely virtual environment, AR integrates digital elements into the user's real-world experience. This is achieved through devices such as smartphones, tablets, AR glasses, and headsets, which use cameras and sensors to detect the physical environment and superimpose digital content accordingly.

Historical Context and Development

The concept of AR dates back to the 1960s, but it has gained significant traction in recent years due to advancements in computing power, sensors, and display technologies. The proliferation of smartphones equipped with high-resolution cameras and powerful processors has made AR more accessible to the general public. Today, AR is used in various fields, including gaming, education, healthcare, and, notably, design and creativity.

8.2.2 Applications of AR in Design and Creativity

Interior and Architectural Design

AR has become an invaluable tool in interior and architectural design. Designers can use AR applications to visualize how furniture, decor, and structural changes will look in a space before making any physical alterations. Apps like IKEA Place allow users to place virtual furniture in their homes to see how it fits and looks. This not only enhances the design process but also helps clients make informed decisions.

Product Design and Prototyping

In product design, AR facilitates rapid prototyping and iteration. Designers can create 3D models of their products and visualize them in real-world settings. This allows for real-time adjustments and improvements, reducing the time and cost associated with traditional prototyping. AR also enables collaborative design processes, where team members can view and manipulate designs remotely.

Fashion and Textile Design

AR is transforming the fashion industry by enabling designers to create virtual try-ons and visualizations. Fashion designers can use AR to project their designs onto mannequins or even live models, allowing for immediate feedback and adjustments.

Brands like Gucci and Burberry have adopted AR for virtual fitting rooms, enhancing the shopping experience by letting customers see how clothes and accessories look on them without trying them on physically.

Graphic and Visual Arts

In graphic design and visual arts, AR provides new dimensions for creativity. Artists can create interactive and dynamic artworks that change based on the viewer's perspective or actions.

AR also allows for the creation of immersive exhibitions and installations, where digital art is integrated into physical spaces. This fusion of digital and physical art opens up new possibilities for storytelling and audience engagement.

Animation and Film Production

AR is increasingly used in animation and film production to create more immersive and interactive experiences. Filmmakers can use AR to visualize special effects and animations in real-time, making the production process more efficient.

AR also enables new forms of storytelling, where viewers can interact with characters and environments in innovative ways. This technology is being explored in augmented reality theaters and interactive film experiences.

8.2.3 Benefits of AR in Design and Creativity

Enhanced Visualization

One of the primary benefits of AR in design and creativity is enhanced visualization. AR allows designers and artists to see their creations in the real world, providing a more accurate representation of how the final product will look and function. This improved visualization helps in identifying potential issues and making necessary adjustments early in the design process.

Increased Engagement and Interactivity

AR creates more engaging and interactive experiences for both creators and audiences. By integrating digital elements into the physical world, AR allows users to interact with designs and artworks in novel ways. This increased interactivity enhances the user experience and can lead to greater engagement and satisfaction. For example, in marketing and advertising, AR campaigns can capture attention and create memorable interactions.

Improved Collaboration and Communication

AR facilitates better collaboration and communication among design teams and stakeholders. Designers can share AR visualizations with clients and team members, enabling them to view and interact with the design in real-time. This collaborative approach helps in aligning visions, gathering feedback, and making informed decisions quickly. Remote collaboration is also enhanced, as AR allows for virtual meetings where participants can view and manipulate designs together.

Cost and Time Efficiency

By enabling rapid prototyping and real-time adjustments, AR helps reduce the time and cost associated with traditional design processes. Physical prototypes and mock-ups can be minimized or eliminated, saving materials and labor. AR also streamlines the approval process, as clients and stakeholders can provide immediate feedback based on accurate visualizations.

8.2.4 Challenges and Considerations

Technical Limitations

Despite its potential, AR still faces technical limitations that can hinder its effectiveness. These include issues with hardware compatibility, software stability, and the accuracy of environmental tracking.

High-quality AR experiences often require advanced devices and robust computing power, which may not be accessible to all users. Overcoming these technical barriers is crucial for broader adoption.

User Experience and Accessibility

Ensuring a seamless and intuitive user experience in AR applications can be challenging. Designers must consider how users will interact with AR content and ensure that the interface is easy to navigate. Accessibility is another concern, as not all users may have the same level of comfort or familiarity with AR technology. Providing clear instructions and user support is essential for successful implementation.

Privacy and Security

The use of AR involves capturing and processing large amounts of data from the physical environment, raising privacy and security concerns. Ensuring that user data is handled responsibly and securely is paramount. Developers must implement robust data protection measures and provide transparency about how data is collected and used. Addressing these concerns is crucial for building trust and ensuring the ethical use of AR technology.

Content Creation and Management

Creating high-quality AR content requires specialized skills and tools, which can be a barrier for some designers and artists. Additionally, managing and updating AR content can be complex, particularly for large-scale projects.

Ensuring that AR experiences are maintained and updated regularly is important for keeping them relevant and functional.

8.2.5 Future Prospects and Innovations

Advances in AR Technology

The future of AR in design and creativity looks promising, with ongoing advancements in technology expected to enhance its capabilities. Improvements in hardware, such as lighter and more powerful AR glasses, will make AR more accessible and comfortable to use. Software developments will lead to more accurate and stable AR experiences, allowing for more complex and detailed visualizations.

Integration with Other Technologies

The integration of AR with other emerging technologies, such as Artificial Intelligence (AI), Internet of Things (IoT), and Virtual Reality (VR), will open up new possibilities for design and creativity. AI can enhance AR experiences by providing intelligent recommendations and automating repetitive tasks.

IoT can enable AR applications to interact with smart objects in the physical environment. Combining AR with VR can create mixed reality experiences that offer even greater immersion and interactivity.

Expanded Applications and Use Cases

As AR technology continues to evolve, its applications in design and creativity will expand. In architecture, AR could be used for virtual site visits and interactive building tours. In education, AR can create immersive learning experiences that make complex concepts more accessible and engaging. The entertainment industry will also continue to explore AR for creating new forms of interactive media and live performances.

Democratization of AR Tools

Efforts to democratize AR tools will make the technology more accessible to a wider range of designers and creatives. User-friendly AR development platforms and tools will enable individuals without specialized skills to create their own AR experiences. This democratization will foster innovation and diversity in the AR content landscape, leading to a richer and more varied array of AR applications.

8.3 FOSTERING INNOVATION VIA HUMAN AUGMENTATION

Human augmentation refers to the use of technology to enhance human physical and cognitive capabilities. This field encompasses a wide range of applications, from prosthetics and exoskeletons to cognitive enhancements and brain-machine interfaces. By pushing the boundaries of what humans can achieve, human augmentation fosters innovation across various domains, including healthcare, industry, education, and everyday life.

Human augmentation is a powerful catalyst for innovation, enhancing physical and cognitive capabilities and opening up new possibilities in various fields. From healthcare and industry to everyday life, these technologies are transforming how we live and work. However, realizing the full potential of human augmentation requires addressing ethical concerns, ensuring safety and accessibility, and promoting equitable access. As technology continues to advance, human augmentation will play a crucial role in shaping a more innovative, inclusive, and capable future.

Human augmentation involves integrating advanced technologies with the human body to enhance physical and mental abilities. This can include wearables, implants, prosthetics, and cognitive enhancers that improve performance, health, and quality of life. The scope of human augmentation is broad, covering medical devices, wearable technology, artificial intelligence, and more.

The concept of enhancing human capabilities is not new. Historical attempts at human augmentation include the use of eyeglasses, hearing aids, and prosthetic limbs. However, recent technological advancements have significantly expanded the potential and sophistication of these augmentations, leading to more integrated and effective solutions.

8.3.1 Applications of Human Augmentation

Medical and Healthcare Innovations

Human augmentation is revolutionizing healthcare by providing advanced solutions for rehabilitation, treatment, and support.

Prosthetics and Bionics

Modern prosthetic limbs equipped with sensors and actuators can mimic natural movements more accurately, improving the lives of amputees. Bionic limbs that connect directly to the nervous system allow users to control their prosthetics with their thoughts, providing a more intuitive and seamless experience.

Exoskeletons

Exoskeletons are wearable devices that support and enhance the wearer's movements. In healthcare, exoskeletons are used for rehabilitation, helping patients regain mobility after spinal cord injuries or strokes. They also assist individuals with mobility impairments in performing daily activities.

Neural Implants

Neural implants, such as cochlear implants for the deaf and deep brain stimulators for Parkinson's disease, demonstrate how technology can restore lost functions. These devices enhance sensory and cognitive functions, improving patients' quality of life and enabling them to engage more fully in society.

Industrial and Workplace Applications

Human augmentation technologies are transforming the workplace by enhancing productivity, safety, and efficiency.

Wearable Technology

Wearable devices, including smart glasses and augmented reality (AR) headsets, provide workers with real-time information and guidance, improving accuracy and efficiency. For instance, AR headsets can overlay schematics and instructions onto the physical world, assisting workers in complex tasks.

Exoskeletons in Industry

In industrial settings, exoskeletons reduce physical strain on workers, preventing injuries and increasing endurance. These devices are particularly useful in manufacturing, construction, and logistics, where heavy lifting and repetitive tasks are common.

Cognitive Augmentation

Cognitive augmentation aims to enhance mental abilities such as memory, attention, and problem-solving.

Nootropics and Cognitive Enhancers

Nootropics, also known as smart drugs, are substances that enhance cognitive functions. These include prescription medications for attention deficit hyperactivity disorder (ADHD) and experimental drugs aimed at improving memory and learning capabilities.

Brain-Computer Interfaces

Brain-computer interfaces (BCIs) enable direct communication between the brain and external devices. BCIs have potential applications in enhancing cognitive functions, providing new ways to interact with technology, and even allowing for mind-controlled devices.

8.3.2 Benefits of Human Augmentation

Enhanced Capabilities

Human augmentation enhances physical and cognitive capabilities, allowing individuals to perform tasks that would otherwise be impossible or difficult. This increased capability can lead to higher productivity, better health outcomes, and improved quality of life.

Innovation and Creativity

By augmenting human abilities, these technologies foster innovation and creativity. Enhanced cognitive functions and new ways of interacting with technology can lead to novel ideas and solutions. In creative fields, augmented reality and brain-computer interfaces open up new possibilities for artistic expression and design.

Inclusivity and Accessibility

Human augmentation promotes inclusivity by providing solutions that enable individuals with disabilities to participate more fully in society. Advanced prosthetics, exoskeletons, and neural implants help bridge the gap between ability and disability, allowing for greater independence and social integration.

Economic Impact

The development and deployment of human augmentation technologies have significant economic implications. These technologies can reduce healthcare costs by improving treatment outcomes and preventing injuries. In the workplace, augmented workers can achieve higher productivity and efficiency, contributing to economic growth.

8.3.3 Challenges and Ethical Considerations

Ethical Concerns

The enhancement of human capabilities raises ethical questions about equity, identity, and the nature of humanity. Issues such as access to augmentation technologies, potential societal divides between augmented and non-augmented individuals, and the impact on human identity and agency need careful consideration.

Health and Safety

Ensuring the safety and effectiveness of human augmentation technologies is crucial. Rigorous testing and regulatory oversight are necessary to prevent potential health risks associated with new devices and treatments.

Long-term effects of cognitive enhancers and neural implants also need thorough investigation.

Privacy and Security

The integration of technology with the human body raises concerns about privacy and security. Augmentation devices that collect and transmit data must be safeguarded against hacking and misuse. Protecting the personal and sensitive information of users is essential to maintain trust and ensure ethical use.

Accessibility and Cost

While human augmentation holds great promise, its benefits must be accessible to all, not just a privileged few. High costs and limited availability can restrict access, exacerbating social inequalities.

Policies and initiatives that promote affordability and widespread adoption are necessary to ensure equitable access.

8.3.4 Future Prospects and Innovations

Advances in Technology

Ongoing advancements in fields such as artificial intelligence, biotechnology, and nanotechnology will continue to drive innovation in human augmentation. AI can enhance the functionality of prosthetics and cognitive enhancers, while biotechnology and nanotechnology offer new possibilities for integrating technology with the human body.

Personalized Augmentation

The future of human augmentation lies in personalization. Tailoring augmentation technologies to individual needs and preferences will maximize their effectiveness and user satisfaction. Personalized medicine, custom prosthetics, and adaptive cognitive enhancers are examples of how this trend is shaping the future.

Integration with Other Technologies

Human augmentation will increasingly integrate with other emerging technologies, such as the Internet of Things (IoT) and blockchain. IoT can enable augmented devices to communicate and collaborate, creating more seamless and interconnected experiences. Blockchain can provide secure and transparent records for the use and regulation of augmentation technologies.

Societal Impact

As human augmentation becomes more prevalent, its impact on society will be profound. These technologies have the potential to transform education, work, and daily life, promoting a more inclusive and capable society. However, managing this transformation requires careful consideration of ethical, social, and economic factors.

CHAPTER 9
ETHICS AND SOCIETAL IMPACT

9.1 ETHICAL CONSIDERATIONS IN HUMAN AUGMENTATION

Human augmentation, the use of technology to enhance physical and cognitive abilities, presents significant ethical challenges. While the potential benefits of augmentation are substantial, including improved health, productivity, and quality of life, they come with complex ethical dilemmas.

Human augmentation holds tremendous promise for enhancing human capabilities and improving quality of life. However, realizing this potential requires careful consideration of the ethical implications. Addressing issues of equity, identity, privacy, and societal impact is crucial to ensuring that these technologies are developed and used responsibly. By establishing robust ethical frameworks, promoting responsible innovation, and engaging the public, we can navigate the ethical challenges of human augmentation and harness its benefits for the betterment of society.

Human augmentation encompasses a wide range of technologies designed to enhance human capabilities. This includes physical enhancements like prosthetics and exoskeletons, cognitive enhancements such as brain-computer interfaces and nootropics, and sensory enhancements like bionic eyes and hearing aids. These technologies promise to improve human performance and overcome limitations, but they also raise profound ethical questions.

Historically, humans have sought to enhance their abilities through various means, from simple tools to sophisticated medical devices. However, recent advances in biotechnology, artificial intelligence, and nanotechnology have significantly expanded the scope and impact of human augmentation. These advancements necessitate a careful examination of the ethical implications.

9.1.1 Equity and Access

Socioeconomic Disparities

One of the primary ethical concerns with human augmentation is the potential to exacerbate existing socioeconomic disparities. Advanced augmentation technologies are often expensive and may be accessible only to the wealthy, creating a divide between those who can afford enhancements and those who cannot. This could lead to a society where enhanced individuals have significant advantages in terms of health, education, and employment opportunities, deepening social inequality.

Global Inequities

The disparity in access to augmentation technologies is not only a domestic issue but also a global one. Developing countries may lack the resources to provide these advanced technologies to their populations, leading to global inequities in health and capability. This could reinforce existing global power imbalances and limit the potential benefits of human augmentation to a small fraction of the world's population.

9.1.2 Identity and Human Nature

Redefining Humanity

Human augmentation challenges traditional notions of what it means to be human. Enhancements that significantly alter physical or cognitive abilities raise questions about identity and the essence of humanity. For instance, if brain-computer interfaces can enhance cognitive functions to superhuman levels, what distinguishes human intelligence from artificial intelligence? These questions force us to reconsider the boundaries of human nature and identity.

Psychological and Social Impacts

The integration of augmentation technologies into human bodies can have profound psychological and social impacts. Individuals with significant enhancements may struggle with their sense of identity and how they relate to others. Society may also develop new forms of discrimination, where "natural" humans are perceived as inferior to those who are augmented. Addressing these psychological and social challenges is crucial to ensuring the ethical use of augmentation technologies.

9.1.3 Privacy and Security

Data Collection and Usage

Human augmentation technologies often involve the collection and processing of large amounts of personal data. For example, brain-computer interfaces and wearable health monitors continuously collect data about an individual's physical and mental states. Ensuring the privacy and security of this data is a significant ethical concern. There is a risk that personal information could be misused by corporations or governments, leading to breaches of privacy and autonomy.

Consent and Autonomy

Obtaining informed consent for the use of augmentation technologies is critical.

Individuals must fully understand the potential risks and benefits of these technologies, as well as how their data will be used and protected. This includes ensuring that individuals are not coerced into using augmentation technologies and that they retain control over their own bodies and data.

9.1.4 Societal Impact

Workforce and Employment

The widespread adoption of human augmentation technologies could transform the workforce and employment landscape. Enhanced individuals may be able to perform tasks more efficiently and effectively than their non-augmented counterparts, leading to potential job displacement and changes in the nature of work. Ethical considerations include ensuring fair labor practices and supporting those who may be adversely affected by these changes.

Ethical Use in Military and Law Enforcement

Human augmentation technologies also have significant implications for military and law enforcement. Enhanced soldiers or officers could have advantages in terms of strength, endurance, and cognitive abilities. However, the use of these technologies in such contexts raises ethical questions about the potential for abuse, the escalation of conflict, and the protection of human rights. Clear guidelines and regulations are necessary to ensure the ethical use of augmentation in these fields.

Long-Term Societal Changes

The integration of human augmentation technologies into society will likely result in long-term changes that need to be carefully managed. These changes could include shifts in social norms, values, and expectations regarding human abilities and performance. Ensuring that these changes are positive and inclusive requires proactive ethical consideration and regulation.

9.1.5 Regulatory and Ethical Frameworks

Establishing Guidelines

To address the ethical challenges of human augmentation, comprehensive guidelines and frameworks must be established. These guidelines should be developed through inclusive and transparent processes that involve stakeholders from diverse backgrounds, including ethicists, scientists, policymakers, and the public.

The aim should be to create regulations that balance innovation with ethical responsibility, ensuring that the benefits of augmentation technologies are accessible and equitable.

International Cooperation

Given the global nature of technology development and the potential for wide-reaching impacts, international cooperation is essential. Establishing global standards and agreements can help ensure that human augmentation technologies are developed and used ethically across different countries and cultures. This includes addressing issues of access, equity, and the protection of human rights on a global scale.

9.1.6 Promoting Ethical Innovation

Responsible Research and Development

Encouraging responsible research and development practices is crucial for the ethical advancement of human augmentation technologies. This involves conducting thorough ethical reviews, considering long-term implications, and prioritizing the well-being of individuals and society. Researchers and developers must be committed to ethical principles and transparency in their work.

Public Engagement and Education

Engaging the public in discussions about human augmentation is essential for building trust and understanding. Public education initiatives can help individuals make informed decisions about the use of augmentation technologies and participate in shaping ethical guidelines. Inclusive dialogue can also ensure that diverse perspectives and values are considered in the development and regulation of these technologies.

Ethical Business Practices

Companies developing and marketing human augmentation technologies have a responsibility to uphold ethical standards. This includes ensuring transparency in product claims, providing accurate information about risks and benefits, and prioritizing the privacy and security of users. Ethical business practices also involve considering the broader social impact of their technologies and contributing to efforts that promote equitable access.

9.2 ADDRESSING PRIVACY AND SECURITY CONCERNS

Privacy and security are paramount concerns in today's digital age, where technology permeates every aspect of our lives. As innovations continue to advance, from artificial intelligence (AI) to biometric authentication, the need to safeguard personal data and ensure secure systems becomes increasingly critical.

Privacy and security concerns in technology are multifaceted and require comprehensive strategies to address effectively. By prioritizing ethical considerations, complying with regulatory frameworks, and adopting best practices, organizations can protect personal data, enhance cybersecurity resilience, and foster trust among users. Continuous innovation and collaboration across industries, governments, and academia are essential to adapt to evolving threats and opportunities in the digital landscape. Upholding privacy rights, safeguarding sensitive information, and promoting secure technologies contribute to a safer and more resilient digital future for individuals and societies worldwide.

9.2.1 Introduction to Privacy and Security in Technology

Definition and Importance

Privacy refers to the right of individuals to control their personal information and how it is collected, used, and shared by others. Security, on the other hand, pertains to protecting systems, networks, and data from unauthorized access, breaches, and cyber threats. Together, privacy and security form the foundation of trust in digital interactions, essential for individuals, organizations, and societies to function and thrive in the digital age.

Evolution and Complexity

The landscape of privacy and security has evolved dramatically with the rapid advancement of technology. From the early days of the internet to today's interconnected world of smart devices and cloud computing, the complexity of managing and protecting personal data has grown exponentially. Emerging technologies such as AI, Internet of Things (IoT), and biometrics present new challenges and opportunities in privacy and security management.

9.2.2 Key Privacy Concerns in Technology

Data Collection and Usage

One of the primary privacy concerns revolves around the collection and usage of personal data by technology companies and service providers. In the digital ecosystem, vast amounts of data are generated through online activities, transactions, and interactions with connected devices. This data can include sensitive information such as financial records, health data, location history, and behavioral patterns.

Surveillance and Monitoring

The proliferation of surveillance technologies, including closed-circuit television (CCTV), facial recognition, and geolocation tracking, raises concerns about individual privacy and civil liberties. Government agencies, law enforcement, and private entities increasingly use these technologies for public safety, security, and marketing purposes. Balancing the benefits of surveillance with the protection of privacy rights remains a contentious issue.

Biometric Data and Identity

Biometric technologies, such as fingerprint scans, iris recognition, and facial biometrics, offer convenient and secure authentication methods. However, the collection and storage of biometric data raise significant privacy concerns. Unlike passwords or PINs, biometric identifiers are inherently tied to individuals and cannot be easily changed if compromised. Ensuring robust security measures and informed consent is essential when handling biometric data.

9.2.3 Security Challenges in Technology

Cyber Threats and Vulnerabilities

Cyber threats, including malware, ransomware, phishing attacks, and data breaches, pose significant security risks to individuals and organizations. Cybercriminals exploit vulnerabilities in software, networks, and human behavior to gain unauthorized access to sensitive data or disrupt operations. The increasing sophistication and frequency of cyber attacks underscore the need for proactive security measures and defense strategies.

Internet of Things (IoT) Security

The IoT, encompassing connected devices ranging from smart home appliances to industrial sensors, presents unique security challenges. Many IoT devices have limited built-in security features and may transmit data over unsecured networks, making them vulnerable to exploitation.

Securing the IoT ecosystem requires implementing robust authentication, encryption, and device management practices to mitigate risks.

Cloud Computing Security

Cloud computing enables scalable storage, processing power, and accessibility to data and applications over the internet. While offering numerous benefits, including cost efficiency and flexibility, cloud services also introduce security considerations.

Organizations must ensure data protection during transmission and storage in the cloud, address compliance requirements, and manage access controls to safeguard sensitive information.

9.2.4 Ethical Considerations in Privacy and Security

Transparency and Accountability

Ethical principles such as transparency and accountability are fundamental to maintaining trust between users, businesses, and governments. Individuals have the right to know how their data is being collected, used, and shared by technology providers. Organizations should implement clear privacy policies, provide meaningful consent mechanisms, and regularly audit their data practices to demonstrate accountability.

Fairness and Non-Discrimination

Fairness and non-discrimination are critical ethical considerations in the use of data-driven technologies, such as AI and machine learning. Algorithms trained on biased data sets can perpetuate biases and lead to unfair outcomes, such as discrimination in hiring, lending, or law enforcement.

Ethical guidelines and diversity in data representation are essential to mitigate biases and ensure equitable treatment.

Respect for User Autonomy

Respecting user autonomy involves empowering individuals to make informed decisions about their privacy preferences and data sharing. It includes offering meaningful consent options, providing accessible privacy settings, and allowing users to opt-out of data collection practices.

Respecting user autonomy also entails minimizing data collection to what is necessary for legitimate purposes and ensuring data accuracy and relevance.

9.2.5 Regulatory Frameworks and Compliance

Global and National Regulations

Regulatory frameworks play a crucial role in addressing privacy and security concerns in technology. Laws such as the European Union's General Data Protection Regulation (GDPR), California Consumer Privacy Act (CCPA), and Brazil's Lei Geral de Proteção de Dados (LGPD) establish guidelines for data protection, user rights, and compliance requirements.

These regulations impose obligations on organizations to implement data protection measures, obtain consent for data processing, and notify individuals of data breaches.

Industry Standards and Best Practices

Industry standards and best practices complement regulatory requirements by providing guidelines for security protocols, data encryption, incident response, and risk management. Standards bodies, industry associations, and cybersecurity organizations develop frameworks such as ISO/IEC 27001 for information security management and NIST Cybersecurity Framework for risk-based cybersecurity practices. Adhering to these standards helps organizations strengthen their security posture and demonstrate commitment to protecting user privacy.

9.2.6 Strategies to Address Privacy and Security Concerns

Data Minimization and Anonymization

Adopting data minimization practices involves collecting only the necessary data for specific purposes and minimizing the retention period. Anonymization techniques can further protect privacy by removing or obfuscating personally identifiable information from datasets used for analysis or research. Implementing these practices reduces the risk of unauthorized access and limits exposure in the event of a data breach.

Encryption and Secure Communication

Encrypting data at rest and in transit ensures that sensitive information remains unreadable to unauthorized parties. Strong encryption algorithms and key management practices safeguard data integrity and confidentiality, whether stored in databases, transmitted over networks, or accessed by authorized users. Secure communication protocols such as Transport Layer Security (TLS) protect data exchanged between devices and servers from interception or tampering.

Multifactor Authentication and Access Controls

Implementing multifactor authentication (MFA) enhances security by requiring multiple forms of verification to access systems or applications. MFA combines passwords or PINs with biometric factors, security tokens, or one-time passcodes to verify user identity.

Access controls enforce least privilege principles, granting permissions based on roles and responsibilities to limit unauthorized access to sensitive data or resources.

Cybersecurity Awareness and Training

Promoting cybersecurity awareness among employees, stakeholders, and the general public is essential for mitigating human errors and improving security hygiene.

Training programs educate individuals on recognizing phishing attempts, practicing safe browsing habits, and reporting suspicious activities. Awareness campaigns raise consciousness about privacy risks, data protection practices, and the importance of maintaining secure behaviors online.

9.2.7 Future Trends and Challenges

Emerging Technologies

The evolution of technology, including AI, quantum computing, and 5G networks, introduces new opportunities and challenges in privacy and security. AI-driven cybersecurity solutions enhance threat detection and response capabilities, while quantum-resistant cryptography prepares for future encryption standards. As 5G enables faster connectivity and IoT proliferation, securing data transmission and protecting edge devices become imperative priorities.

Ethical AI and Algorithmic Accountability

Advancing ethical AI frameworks and promoting algorithmic accountability are critical for addressing biases, ensuring fairness, and building trust in AI-driven systems. Techniques such as explainable AI (XAI) enable transparency by providing insights into how algorithms make decisions. Ethical guidelines emphasize fairness, accountability, and user-centric design principles to mitigate unintended consequences and promote responsible AI deployment.

9.3 SOCIETAL PERCEPTIONS AND CULTURAL IMPACTS

Societal perceptions and cultural impacts are crucial lenses through which we view and understand technological advancements, societal changes, and cultural shifts. In the context of emerging technologies and global interconnectedness, examining how societies perceive and respond to innovations offers valuable insights into human behavior, values, and adaptation processes.

The interplay between societal perceptions, cultural impacts, and technological advancements is complex and dynamic, shaping the future trajectory of societies worldwide. By understanding how perceptions evolve, cultural values adapt, and technologies influence everyday life, stakeholders can navigate challenges, harness opportunities, and foster inclusive, ethical, and sustainable technological advancements. Embracing a multidisciplinary approach that integrates insights from sociology, anthropology, psychology, and technology studies is essential for promoting informed decision-making and shaping a positive future for all.

Understanding Societal Perceptions

Definition and Scope

Societal perceptions encompass collective attitudes, beliefs, and opinions held by communities or populations towards specific issues, phenomena, or changes. These perceptions are shaped by cultural norms, historical contexts, educational systems, media influences, and personal experiences. Understanding societal perceptions provides a foundation for predicting responses to innovations and navigating societal acceptance or resistance.

Factors Influencing Perceptions

Several factors influence societal perceptions, including:

- **Media Representation:** How technologies are portrayed in mainstream media, news outlets, and entertainment influences public opinion.

- **Educational Discourse:** The role of education in shaping perceptions through curricula, academic research, and public discourse.
- **Cultural Values:** Cultural norms, traditions, and values that dictate societal norms and behaviors.
- **Political and Economic Contexts:** Socioeconomic factors and political climates influence perceptions of technological advancements and their impacts on society.

Cultural Impacts of Technological Advancements

Definition and Examples

Cultural impacts refer to the ways in which technological advancements reshape cultural practices, values, traditions, and societal structures. These impacts can be profound, altering how individuals interact, communicate, work, and live.

Examples of Cultural Impacts

- **Communication:** The advent of social media and digital communication platforms has revolutionized how individuals connect and communicate globally, influencing language use, social etiquette, and interpersonal relationships.
- **Work and Employment:** Automation and artificial intelligence are transforming labor markets, leading to debates about job displacement, skills retraining, and the future of work cultures.
- **Arts and Entertainment:** Technologies such as virtual reality (VR), augmented reality (AR), and digital media have revolutionized artistic expression, storytelling, and entertainment experiences.

Societal Perceptions of Emerging Technologies

Case Studies

Exploring societal perceptions towards specific emerging technologies provides insights into acceptance, concerns, and regulatory responses.

- **Artificial Intelligence (AI):** Public perceptions of AI range from excitement about potential benefits in healthcare and autonomous vehicles to concerns about job displacement, ethical implications, and privacy risks.
- **Biotechnology:** Genetic engineering and biotechnological advancements provoke debates on bioethics, genetically modified organisms (GMOs), and personalized medicine.
- **Blockchain and Cryptocurrencies:** Cultural perceptions of blockchain

technologies vary globally, influencing adoption rates, regulatory approaches, and economic implications.

Implications for Policy and Governance

Regulatory Challenges

Addressing societal perceptions and cultural impacts requires proactive policy frameworks and governance structures that balance innovation with ethical considerations, public safety, and societal well-being.

Ethical Considerations

Navigating ethical dilemmas associated with technological advancements, such as privacy concerns, data security, algorithmic biases, and social inequalities.

International Cooperation

Collaborative efforts among governments, industries, academia, and civil society to establish global standards, regulations, and ethical guidelines for emerging technologies.

Addressing Resistance and Promoting Acceptance

Strategies

Implementing strategies to address public skepticism, misinformation, and fears surrounding technological advancements through education, public engagement, transparency, and accountability.

Case Studies

Examining successful strategies and initiatives that promote public acceptance and trust in innovations, highlighting lessons learned and best practices.

CHAPTER 10

HUMAN AUGMENTATION IN HEALTHCARE

10.1 AUGMENTED SURGERY AND MEDICAL IMAGING

In recent years, the convergence of advanced technologies and medical science has ushered in a new era in surgical procedures and medical imaging. Augmented reality (AR) and its integration into surgical practices have transformed the way surgeons visualize, plan, and execute complex procedures. Similarly, advancements in medical imaging techniques, such as MRI, CT scans, and ultrasound, have vastly improved diagnostic accuracy and treatment planning.

Augmented surgery and medical imaging represent transformative innovations that are reshaping the landscape of modern healthcare. From enhancing surgical precision and diagnostic accuracy to improving patient outcomes and advancing medical education, these technologies hold immense promise for the future. However, their successful integration requires addressing challenges related to ethics, accessibility, and technological reliability. By harnessing the synergies between augmented reality and advanced imaging modalities, healthcare providers can usher in a new era of personalized, efficient, and patient-centered care.

By staying at the forcfront of technological advancements and fostering collaboration across disciplines, the healthcare industry can leverage augmented surgery and medical imaging to empower clinicians, improve clinical outcomes, and ultimately enhance the quality of life for patients worldwide.

10.1.1 Technological Advancements in Augmented Surgery

Augmented surgery leverages AR to overlay digital information onto the surgeon's field of view, providing real-time guidance during procedures. One of the pioneering applications is the use of AR in orthopedic surgery, where surgeons can visualize 3D models of bones and joints overlaid onto the patient's anatomy.

This technology enables precise placement of implants and more accurate alignment, leading to improved patient outcomes and faster recovery times. In neurosurgery, AR allows surgeons to navigate complex brain structures with enhanced precision, minimizing damage to surrounding tissues. By integrating pre-operative imaging data into the surgical field, AR systems help surgeons make informed decisions in real-time, reducing the risk of complications and optimizing surgical interventions.

10.1.2 Enhanced Medical Imaging Techniques

Medical imaging plays a crucial role in diagnosis, treatment planning, and post-operative assessment. Recent advancements have significantly enhanced imaging modalities such as MRI, CT scans, and ultrasound. MRI, for instance, provides detailed soft tissue contrast without ionizing radiation, making it invaluable in detecting abnormalities in the brain, spine, and musculoskeletal system. CT scans offer rapid imaging of bony structures and are essential for trauma assessment and surgical planning. Meanwhile, ultrasound continues to evolve with portable devices enabling real-time imaging at the point of care, facilitating guided procedures and interventions.

10.1.3 Integration of Augmented Reality with Medical Imaging

The integration of AR with medical imaging represents a paradigm shift in healthcare delivery. By combining real-time imaging data with AR overlays, surgeons can visualize critical structures, navigate complex anatomies, and perform minimally invasive procedures with unprecedented accuracy. This synergy between AR and medical imaging not only enhances surgical precision but also improves patient safety and reduces operative times. For example, during laparoscopic surgeries, AR overlays can guide surgeons to target lesions precisely, minimizing healthy tissue damage. Similarly, in cardiology, AR assists in navigating catheters through intricate cardiac pathways, improving the efficacy of interventions such as ablation procedures.

10.1.4 Benefits of Augmented Surgery and Medical Imaging

The benefits of augmented surgery and advanced medical imaging are multifaceted. Patients experience reduced recovery times, decreased risk of complications, and improved overall outcomes. Surgeons benefit from enhanced visualization and real-time feedback, allowing for more precise interventions and personalized treatment plans. From a healthcare system perspective, these technologies contribute to cost-effectiveness by optimizing resource utilization and reducing hospital stays. Moreover, medical training and education are revolutionized as AR enables simulated practice scenarios and immersive learning experiences for aspiring surgeons and medical professionals.

10.1.5 Challenges and Ethical Considerations

Despite the promising advancements, augmented surgery and medical imaging present challenges and ethical considerations. Issues such as data privacy, the accuracy of augmented overlays, and the potential for over-reliance on technology require careful consideration. Ensuring patient consent, safeguarding sensitive medical information, and maintaining the integrity of diagnostic algorithms are paramount in the adoption of these technologies. Furthermore, disparities in access to advanced healthcare technologies highlight the need for equitable distribution and affordability. Addressing these challenges requires collaboration among healthcare providers, technology developers, policymakers, and regulatory bodies to establish guidelines and standards that prioritize patient safety and ethical practices.

10.1.6 Future Directions and Innovations

Looking ahead, the future of augmented surgery and medical imaging holds exciting possibilities. Continued advancements in AI and machine learning promise to further enhance the accuracy and efficiency of diagnostic imaging and surgical interventions. Integration with robotics could enable autonomous procedures guided by real-time imaging data, further reducing human error and improving surgical outcomes. Moreover, personalized medicine driven by genetic and molecular imaging techniques may revolutionize treatment strategies, offering tailored therapies based on individual patient profiles. As these technologies evolve, interdisciplinary collaboration and ongoing research will be crucial in unlocking their full potential and addressing emerging healthcare challenges.

10.2 ASSISTIVE TECHNOLOGIES FOR DISABILITIES

Assistive technologies have revolutionized the lives of people with disabilities by providing tools and devices that enhance independence, mobility, communication, and overall quality of life. These technologies span a wide range of innovations, from simple aids like wheelchairs and hearing aids to advanced robotics and smart devices.

Assistive technologies play a pivotal role in empowering individuals with disabilities to lead more independent, fulfilling lives. Through continuous innovation, research, and advocacy, these technologies are breaking down barriers and expanding possibilities for inclusion and equality. Addressing challenges such as affordability, accessibility, and technological integration is essential to maximizing the potential of assistive technologies and ensuring they reach those who need them most.

As we look towards the future, collaboration among stakeholders—including healthcare professionals, technology developers, policymakers, and advocacy groups—will be crucial in driving progress and advancing the field of assistive technologies. By prioritizing inclusivity, innovation, and ethical considerations, we can create a more accessible and equitable world where everyone can thrive, regardless of ability or disability.

10.2.1 Evolution and Types of Assistive Technologies

The history of assistive technologies dates back centuries, with early innovations focused on mobility aids such as canes and crutches. The advent of electricity and modern engineering has significantly expanded the capabilities of assistive devices, leading to the development of powered wheelchairs, prosthetic limbs, and communication aids.

Today, assistive technologies are categorized into several types:

- **Mobility Aids:** Wheelchairs, scooters, and exoskeletons enhance mobility for individuals with physical disabilities.
- **Sensory Aids:** Hearing aids, cochlear implants, and visual aids (like magnifiers or screen readers) assist individuals with sensory impairments.
- **Communication Aids:** Speech-generating devices, Braille displays, and alternative communication methods enable individuals with speech or language disabilities to communicate effectively.
- **Prosthetics and Orthotics:** Artificial limbs, braces, and orthotic devices restore functionality and support for individuals with limb differences or injuries.

10.2.2 Benefits of Assistive Technologies

Assistive technologies offer numerous benefits to individuals with disabilities, empowering them to participate more fully in everyday activities and society. For instance, powered wheelchairs enable independent mobility, enhancing access to education, employment, and social interactions. Similarly, hearing aids and cochlear implants improve communication and quality of life for individuals with hearing loss, enabling them to engage more effectively in conversations and social settings. Communication aids like speech-generating devices provide a voice to those who are non-verbal, fostering independence and self-expression. Prosthetic limbs and orthotic devices restore mobility and function, allowing individuals to perform tasks that would otherwise be challenging or impossible. Overall, these technologies promote inclusion, dignity, and autonomy for people with disabilities.

10.2.3 Challenges and Barriers

Despite their benefits, assistive technologies face several challenges that hinder their widespread adoption and effectiveness. One significant barrier is cost, as many advanced devices are expensive and may not be covered by insurance or accessible to those with limited financial resources. Accessibility concerns also arise, particularly in terms of design and usability for individuals with diverse disabilities. Moreover, technological complexity and interoperability issues can present challenges in integrating different assistive devices into a cohesive system that meets individual needs. Maintenance and repair services are critical for ensuring the longevity and reliability of assistive technologies, yet access to these services may be limited in certain regions or healthcare systems.

10.2.4 Ethical Considerations and Accessibility

Ensuring equitable access to assistive technologies is an ethical imperative. The United Nations Convention on the Rights of Persons with Disabilities emphasizes the right of individuals with disabilities to access assistive devices on an equal basis with others. This includes promoting research and development of affordable technologies, fostering innovation in accessibility features, and advocating for policies that support universal design principles. Furthermore, data privacy and security are crucial considerations, particularly for assistive technologies that collect and transmit personal health information. Safeguarding user data and ensuring informed consent are essential to maintaining trust and protecting the rights of individuals who rely on these technologies.

10.2.5 Technological Innovations and Future Directions

The future of assistive technologies holds promise with ongoing advancements in robotics, artificial intelligence (AI), and wearable devices. Robotics, for example, is enabling more sophisticated prosthetic limbs and exoskeletons that mimic natural movement and enhance functionality. AI-powered devices can adapt to user preferences and environments, providing personalized assistance and predictive capabilities. Wearable technologies such as smart glasses and haptic feedback devices are expanding possibilities for sensory augmentation and navigation assistance. Moreover, the Internet of Things (IoT) is facilitating interconnected assistive devices that can communicate and coordinate tasks seamlessly, further enhancing user independence and convenience.

10.2.6 Impact on Healthcare and Society

Assistive technologies not only benefit individuals with disabilities but also have broader implications for healthcare and society. By enabling greater independence and participation, these technologies reduce dependence on caregivers, alleviate healthcare burdens, and enhance overall well-being. Improved access to education and employment opportunities can contribute to economic empowerment and social inclusion for individuals with disabilities.

From a societal perspective, promoting universal design principles in product development and infrastructure can create more inclusive environments that accommodate diverse needs. Moreover, raising awareness and reducing stigma surrounding disabilities can foster empathy and understanding, promoting a more inclusive and supportive society for all.

10.3 REHABILITATION AND NEURORECOVERY

Rehabilitation and neurorecovery encompass a diverse range of therapies and interventions aimed at restoring function, improving quality of life, and promoting independence for individuals with neurological conditions and injuries.

From stroke rehabilitation to spinal cord injury recovery, these fields integrate medical, therapeutic, and technological approaches to facilitate neurological recovery.

Rehabilitation and neurorecovery represent dynamic fields at the intersection of medicine, technology, and patient-centered care. By harnessing the principles of neuroplasticity and leveraging technological advancements, healthcare providers can empower individuals to achieve meaningful recovery, regain independence, and improve their quality of life. Addressing challenges in accessibility, affordability, and personalized care requires a collaborative effort to advance research, advocate for policies that support equitable access to care, and integrate innovative therapies into clinical practice.

As we look towards the future, continued innovation and interdisciplinary collaboration will drive progress in rehabilitation and neurorecovery, paving the way for enhanced treatment options and improved outcomes for individuals with neurological conditions and injuries. By prioritizing patient-centered approaches and ethical considerations, we can shape a more inclusive healthcare landscape that supports the diverse needs and aspirations of individuals on their journey towards recovery and rehabilitation.

10.3.1 Principles of Rehabilitation and Neurorecovery

The core principles of rehabilitation and neurorecovery revolve around neuroplasticity—the brain's ability to reorganize and form new neural connections in response to injury or disease. Rehabilitation programs are tailored to stimulate neuroplasticity through repetitive exercises, task-specific training, and sensory stimulation. The goal is to promote recovery of motor and cognitive functions, maximize independence, and enhance overall quality of life. Multidisciplinary teams comprising physiatrists, physical therapists, occupational therapists, speech-language pathologists, psychologists, and social workers collaborate to create personalized treatment plans. These plans address the unique needs and goals of each patient, integrating evidence-based practices and innovative technologies to optimize outcomes.

10.3.2 Methods and Approaches in Rehabilitation

Rehabilitation strategies vary depending on the type and severity of the neurological condition or injury. Common approaches include:

- **Physical Therapy**: Focuses on improving mobility, strength, balance, and coordination through exercises, manual techniques, and assistive devices.
- **Occupational Therapy**: Aims to enhance daily living skills, such as self-care, work tasks, and leisure activities, by adapting environments and teaching compensatory strategies.
- **Speech-Language Therapy**: Addresses communication disorders and swallowing difficulties through exercises, communication devices, and techniques to improve speech, language, and swallowing function.
- **Cognitive Rehabilitation**: Targets cognitive impairments such as memory deficits, attention problems, and executive function challenges through structured exercises and cognitive training programs.
- **Electrical Stimulation**: Utilizes electrical currents to stimulate nerves and muscles, promoting muscle re-education, pain relief, and functional recovery.

10.3.3 Technological Advancements in Rehabilitation

Recent technological advancements have revolutionized rehabilitation and neurorecovery by enhancing treatment options and improving patient outcomes. Examples include:

- **Robot-Assisted Therapy**: Robots assist patients in performing repetitive movements with precision, promoting motor recovery after stroke or spinal cord injury.
- **Virtual Reality (VR) and Augmented Reality (AR)**: Immersive environments and interactive simulations provide engaging rehabilitation exercises, facilitating motor learning and sensory integration.
- **Brain-Computer Interfaces (BCIs)**: Enable direct communication between the brain and external devices, allowing individuals with severe disabilities to control assistive technologies or prosthetic limbs through neural signals.
- **Exoskeletons**: Wearable robotic devices support and enhance mobility for individuals with gait impairments, enabling upright walking and reducing reliance on mobility aids.
- **Neurostimulation**: Techniques such as transcranial magnetic stimulation (TMS) and deep brain stimulation (DBS) modulate neural activity, potentially improving motor function and cognitive abilities.

10.3.4 Challenges in Rehabilitation and Neurorecovery

Despite technological advancements, rehabilitation and neurorecovery face several challenges that impact treatment efficacy and accessibility. Limited access to specialized rehabilitation services, particularly in rural or underserved areas, can delay initiation of therapy and hinder recovery outcomes.

Financial barriers, including insurance coverage limitations and high out-of-pocket costs for assistive devices, may prevent individuals from accessing necessary treatments and technologies. Furthermore, variability in patient response to therapies underscores the need for personalized treatment plans tailored to individual needs and goals. Interdisciplinary collaboration among healthcare providers is essential to address complex medical and psychosocial factors influencing rehabilitation outcomes.

Additionally, navigating regulatory frameworks and ensuring safety and efficacy of emerging technologies require ongoing research and development.

10.3.5 Ethical Considerations in Rehabilitation

Ethical considerations in rehabilitation and neurorecovery encompass issues related to patient autonomy, informed consent, and equitable access to care. Respecting patient preferences and goals is paramount in developing collaborative treatment plans that align with individual values and cultural beliefs.

Informed consent ensures that patients are fully informed about potential risks, benefits, and alternatives to treatment, empowering them to make informed decisions about their care. Equity in access to rehabilitation services and technologies is critical to addressing disparities based on socioeconomic status, geographic location, and disability type.

Advocating for policies that promote inclusivity, support research funding, and improve healthcare infrastructure can enhance accessibility and reduce barriers to optimal rehabilitation outcomes.

10.3.6 Future Directions and Innovations

The future of rehabilitation and neurorecovery holds promise with ongoing research and technological innovations aimed at enhancing treatment efficacy and patient engagement.

Advances in neuroimaging techniques, such as functional MRI and diffusion tensor imaging, provide insights into neural networks and recovery mechanisms, guiding personalized treatment approaches.

Integration of AI and machine learning algorithms enables data-driven decision-making in rehabilitation, predicting patient outcomes and optimizing therapy protocols.

Wearable sensors and mobile health applications facilitate remote monitoring of patient progress, promoting continuity of care and enabling real-time adjustments to treatment plans.

Moreover, regenerative medicine and stem cell therapies offer potential avenues for neural repair and regeneration, potentially reversing the effects of neurological damage and enhancing long-term recovery outcomes. Collaborative efforts among researchers, clinicians, industry partners, and advocacy groups are essential to translating these innovations into clinical practice and improving outcomes for individuals with neurological conditions and injuries.

CHAPTER 11

HUMAN AUGMENTATION IN SPORTS AND PERFORMANCE

11.1 PERFORMANCE ENHANCEMENT IN ATHLETICS

Performance enhancement in athletics involves the pursuit of improving athletes' physical and mental capabilities to achieve superior athletic achievements. This field encompasses a wide range of strategies, including training methodologies, nutritional interventions, sports psychology techniques, and advancements in sports science and technology.

Performance enhancement in athletics is a dynamic and multifaceted field that integrates training methodologies, nutritional strategies, sports psychology, and technological innovations to optimize athletic performance. By leveraging scientific advancements and ethical principles, athletes can achieve their full potential while upholding the integrity and fairness of sport.

As the landscape of sports continues to evolve, ongoing research, innovation, and collaboration among athletes, coaches, sports scientists, and policymakers will drive progress in performance enhancement.

Embracing holistic approaches to training, addressing challenges, and embracing ethical considerations will shape the future of athletic excellence and inspire the next generation of athletes worldwide.

11.1.1 Training Methodologies and Techniques

Athletic performance enhancement begins with comprehensive training methodologies designed to optimize strength, speed, agility, endurance, and skill development. Coaches and sports scientists employ periodization techniques to structure training programs into phases that focus on different aspects of fitness and performance. This approach helps prevent overtraining, promote recovery, and maximize adaptations to training stimuli.

Specific training techniques include:

- **Strength Training**: Emphasizes resistance exercises to increase muscle mass, strength, and power, essential for sports requiring explosive movements like sprinting and weightlifting.

- **Speed and Agility Drills:** Improve reaction time, acceleration, and change of direction capabilities through sprint intervals, ladder drills, and cone exercises.
- **Endurance Training:** Enhances cardiovascular fitness and stamina through long-distance running, cycling, swimming, or high-intensity interval training (HIIT).
- **Skill-Specific Training:** Focuses on refining technical skills and tactical understanding relevant to the athlete's sport, such as ball control in soccer or shooting accuracy in basketball.

11.1.2 Nutritional Strategies for Performance Enhancement

Nutrition plays a critical role in optimizing athletic performance by fueling workouts, supporting recovery, and promoting overall health. Athletes adhere to personalized nutrition plans tailored to their training demands, body composition goals, and competition schedule. Key nutritional strategies include:

- **Macronutrient Balance:** Adjusting carbohydrate, protein, and fat intake to meet energy needs and support muscle repair and growth.
- **Hydration:** Maintaining fluid balance to prevent dehydration and optimize performance, especially in endurance sports and hot environments.
- **Supplementation:** Using supplements like creatine, caffeine, and electrolytes to enhance energy production, delay fatigue, and support recovery.
- **Periodic Fasting:** Utilizing intermittent fasting or strategic nutrient timing to improve metabolic flexibility and potentially enhance performance adaptations.

11.1.3 Sports Psychology and Mental Conditioning

Sports psychology techniques focus on enhancing mental resilience, concentration, confidence, and motivation to optimize athletic performance. Psychologists work with athletes to develop mental skills such as goal-setting, visualization, relaxation techniques, and mindfulness practices.

These strategies help athletes manage performance anxiety, overcome setbacks, and maintain focus during competition. Cognitive-behavioral interventions address negative thought patterns and promote positive self-talk, fostering a mindset conducive to peak performance. Mental conditioning also includes techniques for stress management, arousal regulation, and psychophysiological training to synchronize mind and body responses under pressure.

11.1.4 Technological Advances in Sports Science

Advancements in sports science and technology have revolutionized performance monitoring, biomechanical analysis, and injury prevention in athletics. Key innovations include:

- **Wearable Technology**: Devices like GPS trackers, heart rate monitors, and accelerometers provide real-time data on athletes' physiological metrics, training loads, and movement patterns.
- **Biomechanical Analysis:** High-speed cameras, motion capture systems, and force plates analyze athletes' movement mechanics to optimize technique, prevent injuries, and enhance efficiency.
- **Virtual Reality (VR) Training**: Immersive simulations enable athletes to practice and refine skills in a controlled environment, enhancing decision-making and spatial awareness.
- **Recovery Technology**: Cryotherapy chambers, compression garments, and recovery boots accelerate muscle recovery, reduce inflammation, and alleviate soreness post-exercise.

11.1.5 Ethical Considerations in Performance Enhancement

Performance enhancement in athletics raises ethical considerations related to fairness, safety, and the integrity of sport. Athletes and coaches must navigate the ethical implications of using performance-enhancing substances (e.g., steroids, stimulants) and methods (e.g., blood doping, gene doping) that may compromise health and violate anti-doping regulations. Balancing competitive pressures with ethical principles requires transparency, informed consent, and adherence to anti-doping policies and regulations. Ensuring fair play and maintaining thc spirit of sport uphold the integrity of athletic competition and safeguard the health and well-being of athletes.

11.1.6 Challenges in Performance Enhancement

Achieving optimal athletic performance is accompanied by various challenges, including injury risk, overtraining, and psychological burnout. Balancing training intensity and recovery periods is crucial to prevent physical and mental fatigue, minimize injury occurrence, and sustain long-term performance gains.

Moreover, navigating the complexities of sports governance, funding disparities, and access to resources can impact athletes' opportunities for development and success. Addressing these challenges requires collaborative efforts among athletes, coaches, sports organizations, and healthcare professionals to promote holistic approaches to performance enhancement.

11.1.7 Future Directions and Innovations

The future of performance enhancement in athletics will likely be shaped by continued advancements in sports science, technology, and interdisciplinary collaboration. Research in genetic profiling, personalized medicine, and regenerative therapies may offer insights into individualized training responses and injury prevention strategies. Integration of AI and machine learning algorithms could optimize training programs, predict injury risks, and enhance performance analytics based on real-time data.

Innovations in wearable technology and biometric sensors may further enhance performance monitoring and recovery strategies, empowering athletes to achieve peak performance while minimizing health risks. Furthermore, ethical frameworks and policy developments will continue to evolve to address emerging challenges and ensure the equitable and responsible application of performance enhancement strategies in sports.

11.2 MONITORING AND TRAINING WITH WEARABLE TECHNOLOGIES

Wearable technologies have revolutionized the way athletes monitor and optimize their training routines. These devices, ranging from fitness trackers to sophisticated biometric sensors, provide real-time data on physiological metrics, movement patterns, and performance metrics.

Wearable technologies are transforming athletic training and performance monitoring by providing real-time insights into athletes' physiological responses, movement patterns, and performance metrics. From enhancing training effectiveness and optimizing performance to preventing injuries and promoting recovery, wearables play a crucial role in modern sports science and coaching practices.

As wearable technologies continue to evolve, addressing challenges related to data accuracy, privacy, user experience, and ethical considerations will be essential. By embracing innovation, leveraging data-driven insights, and upholding ethical standards, athletes, coaches, and sports scientists can harness the full potential of wearable technologies to achieve peak performance and advance the field of sports training and performance optimization.

11.2.1 Evolution of Wearable Technologies in Sports

The evolution of wearable technologies in sports can be traced back to basic fitness trackers that monitored steps and heart rate. Over time, advancements in sensor technology, miniaturization, and data analytics have enabled the development of more sophisticated wearables capable of measuring a wide range of physiological parameters. Today, athletes can utilize wearable devices to monitor not only basic metrics like heart rate and steps but also advanced metrics such as oxygen saturation, muscle activity, and sleep patterns.

11.2.2 Benefits of Wearable Technologies in Athletic Training

Wearable technologies offer numerous benefits to athletes and coaches, enhancing training effectiveness, performance monitoring, and injury prevention:

- **Real-Time Feedback**: Athletes receive immediate feedback on their training intensity, heart rate zones, and performance metrics, enabling them to adjust their efforts in real-time to optimize training sessions.
- **Objective Monitoring**: Objective data from wearables provide insights into athletes' physiological responses, movement mechanics, and recovery patterns, facilitating evidence-based decision-making in training and competition.
- **Performance Optimization:** Wearables help identify trends and patterns in athletes' performance, allowing coaches to tailor training programs, optimize recovery strategies, and maximize athletic potential.
- **Injury Prevention**: Monitoring biomechanical metrics and workload helps identify injury risks early, allowing for proactive interventions and adjustments to training loads to minimize the risk of overuse injuries.

11.2.3 Applications of Wearable Technologies in Athletic Training

Wearable technologies find diverse applications across various sports and training settings:

- **Endurance Sports:** GPS-enabled wearables track distance, speed, and route maps for runners, cyclists, and triathletes, providing insights into pacing strategies and performance improvements.
- **Strength Training**: Wearables measure muscle activity, power output, and movement velocity during resistance exercises, guiding athletes in optimizing workout intensity and technique.
- **Team Sports**: Athletes in team sports wear GPS trackers and accelerometers to monitor movement patterns, sprint distances, and workload metrics, informing coaches about players' physical exertion and recovery needs.
- **Rehabilitation**: Wearables aid in monitoring recovery progress and adherence to rehabilitation protocols following injuries, facilitating safe return-to-play decisions for athletes.

11.2.4 Challenges in Adopting Wearable Technologies

Despite their benefits, wearable technologies in sports face several challenges that impact adoption and integration:

- **Data Accuracy and Reliability:** Ensuring the accuracy and reliability of wearable data is critical for making informed decisions in training and competition.
- **Privacy and Data Security:** Protecting athletes' personal health data and ensuring compliance with data privacy regulations are paramount concerns in wearable technology usage.
- **User Interface and Experience:** Designing intuitive interfaces and user-friendly experiences for athletes and coaches to interpret and utilize wearable data effectively.
- **Integration with Coaching Practices:** Bridging the gap between wearable data insights and coaching practices requires collaboration and communication between sports scientists, coaches, and athletes.

11.2.5 Future Directions and Innovations in Wearable Technologies

The future of wearable technologies in sports holds promise with ongoing innovations and advancements:

- **Miniaturization and Wearability:** Continued miniaturization of sensors and improvements in wearable design will enhance comfort, durability, and user acceptance.
- **Artificial Intelligence and Machine Learning:** Integration of AI algorithms will enable predictive analytics, personalized training recommendations, and real-time performance insights based on wearable data.
- **Biometric Integration:** Wearables may incorporate additional biometric measurements such as blood glucose levels, lactate thresholds, and hormonal responses to provide a comprehensive picture of athlete health and performance.
- **Smart Fabrics and Biofeedback:** Development of smart textiles and biofeedback mechanisms that monitor physiological parameters through clothing or embedded sensors.

11.2.6 Ethical Considerations in Wearable Technology Use

Ethical considerations in wearable technology use include data privacy, informed consent, and the responsible use of athlete data. Athletes must have control over their personal health information collected by wearables, with clear guidelines on data storage, sharing, and usage. Moreover, ensuring transparency in how wearable data influences training decisions and performance evaluations promotes trust between athletes, coaches, and sports organizations. Upholding ethical principles ensures that wearable technologies contribute positively to athlete well-being, performance enhancement, and the integrity of sport.

11.3 BALANCING FAIRNESS AND INNOVATION IN SPORTS

The intersection of fairness and innovation in sports represents a delicate balance between promoting technological advancements and maintaining the integrity of competition. As sports evolve with new technologies, equipment, and training methodologies, questions arise about their impact on fairness, athlete safety, and the spirit of sport.

Balancing fairness and innovation in sports requires thoughtful consideration of ethical principles, regulatory frameworks, and the impact of technological advancements on athletes, competition integrity, and societal values. Embracing innovation while upholding the principles of fair play, inclusivity, and ethical conduct ensures that sports continue to inspire, entertain, and unite people worldwide. By fostering dialogue, collaboration, and responsible governance, stakeholders can navigate the complexities of sports innovation to create a sustainable and equitable future for athletics in the 21st century.

11.3.1 The Role of Technology in Sports Innovation

Technology has profoundly influenced sports innovation, offering tools and advancements that enhance training, performance analysis, and spectator engagement. Innovations range from wearable sensors and biometric tracking devices to advanced materials in equipment and digital platforms for coaching and fan interaction. These technologies provide athletes with insights into their performance metrics, facilitate data-driven coaching decisions, and offer fans immersive viewing experiences.

11.3.2 Enhancing Performance vs. Unfair Advantages

While technological innovations aim to enhance athletic performance and optimize training methods, concerns about unfair advantages and competitive integrity arise. Innovations such as high-performance gear, aerodynamic suits in swimming and track events, and advanced training techniques may create disparities between athletes who have access to these resources and those who do not. Moreover, performance-enhancing substances and methods present ethical dilemmas, as athletes seek ways to gain a competitive edge within the boundaries of anti-doping regulations. Balancing the pursuit of excellence with ethical considerations is crucial to upholding the principles of fair play, sportsmanship, and equal opportunities for all competitors.

11.3.3 Regulatory Frameworks and Ethics in Sports

Governance and regulatory bodies in sports play a pivotal role in establishing guidelines, rules, and ethical standards to govern technological innovations and maintain fairness. Organizations such as the International Olympic Committee (IOC), World Anti-Doping Agency (WADA), and various sports federations continuously update policies to address emerging technologies and ensure a level playing field.

Ethical considerations include:

- **Data Privacy:** Protecting athlete data collected through technologies from misuse or unauthorized access.
- **Informed Consent:** Ensuring athletes understand the implications of using new technologies and performance-enhancing methods.
- **Equitable Access:** Promoting access to innovations without compromising fairness or creating financial barriers for athletes from less affluent backgrounds.

11.3.4 Benefits of Innovation in Sports

Innovation in sports offers several benefits that contribute to athlete development, spectator engagement, and the evolution of sports as a whole:

- **Performance Enhancement:** Technologies improve training efficiency, optimize recovery, and enhance biomechanical analysis, allowing athletes to reach their full potential.
- **Safety and Injury Prevention:** Innovations in protective gear and training techniques reduce injury risks and promote long-term athlete health.
- **Fan Experience:** Digital advancements in broadcasting, virtual reality (VR), and interactive media platforms enhance fan engagement and broaden the reach of sports events globally.

11.3.5 Challenges in Balancing Fairness and Innovation

Achieving a balance between fostering innovation and maintaining fairness in sports presents several challenges:

- **Technological Advancements:** Rapid advancements in technology outpace regulatory frameworks, creating dilemmas in governing new innovations.
- **Inequality:** Disparities in access to advanced equipment and training methods may widen the gap between elite athletes and those with limited resources.

- **Ethical Dilemmas:** Decisions about permissible enhancements and the boundaries of fair play are complex and require ongoing ethical debates and consensus-building.

11.3.6 Ethical Considerations in Sports Innovation

Ethical considerations in sports innovation encompass:

- **Integrity of Competition:** Ensuring fairness and preserving the authenticity of athletic achievements.
- **Health and Safety:** Balancing performance gains with athlete well-being and injury prevention.
- **Long-Term Impacts:** Assessing the societal and cultural implications of technological advancements in sports.

11.3.7 Future Directions in Balancing Fairness and Innovation

Looking ahead, the future of sports innovation will be shaped by:

- **Advancements in Biotechnology:** Including gene editing, regenerative medicine, and personalized performance enhancements.
- **AI and Data Analytics:** Enhancing predictive capabilities, performance monitoring, and training optimization.
- **Sustainability:** Developing eco-friendly sports technologies and promoting environmental stewardship.

CHAPTER 12

FUTURE TRENDS AND EMERGING TECHNOLOGIES

12.1 INTEGRATING AI AND HUMAN AUGMENTATION

The integration of Artificial Intelligence (AI) and human augmentation represents a transformative frontier where technology enhances human capabilities across various domains. From healthcare and industry to daily life, AI-driven augmentation offers potential advancements in productivity, decision-making, and quality of life.

The integration of AI and human augmentation represents a frontier of innovation with profound implications for healthcare, industry, education, and daily life. By harnessing AI's capabilities to enhance human abilities responsibly, we can unlock new possibilities for improving productivity, accessibility, and quality of life. Addressing ethical challenges, regulatory frameworks, and societal implications will be crucial in shaping a future where AI-driven augmentation benefits individuals and society equitably and ethically. As technologies evolve, fostering dialogue, collaboration, and ethical stewardship will be essential in realizing the full potential of AI and human augmentation for the betterment of humanity.

12.1.1 Understanding Human Augmentation

Human augmentation involves enhancing human abilities through technological interventions, ranging from prosthetics and wearable devices to neural implants and cognitive enhancements. These technologies aim to improve physical, cognitive, sensory, or emotional functions, enabling individuals to surpass natural limitations and achieve enhanced performance and well-being.

12.1.2 Role of Artificial Intelligence in Human Augmentation

Artificial Intelligence plays a pivotal role in human augmentation by enabling smart, adaptive systems that enhance and extend human capabilities. Key applications include:

- **Prosthetics and Robotics:** AI-powered prosthetic limbs and exoskeletons restore mobility and dexterity, adapting to user movements and environments.
- **Neural Interfaces:** Brain-computer interfaces (BCIs) translate neural

signals into commands for external devices, enabling individuals with paralysis to control computers or robotic limbs.

- **Cognitive Enhancement**: AI-driven cognitive aids and neurostimulation techniques enhance memory, attention, and decision-making abilities, benefiting individuals with cognitive impairments or neurodegenerative diseases.

12.1.3 Applications of AI and Human Augmentation

The integration of AI and human augmentation finds diverse applications across sectors:

- **Healthcare:** AI-enhanced diagnostics, robotic surgeries, and personalized medicine improve patient outcomes and surgical precision.
- **Industry**: AI-powered exoskeletons and augmented reality (AR) tools enhance worker safety, efficiency, and productivity in manufacturing and hazardous environments.
- **Education:** Adaptive learning platforms and AI tutors personalize educational experiences, catering to individual learning styles and abilities.
- **Entertainment:** Virtual reality (VR) and AI-driven gaming experiences create immersive simulations and interactive entertainment.

12.1.4 Ethical Considerations in AI and Human Augmentation

Ethical considerations in AI-driven human augmentation encompass privacy, autonomy, equality, and the potential for misuse or unintended consequences. Key issues include:

- **Privacy and Data Security:** Safeguarding personal health data and neural information collected by AI-enhanced devices.
- **Autonomy and Informed Consent:** Ensuring individuals have control over augmentation choices and understand risks and benefits.
- **Equity and Accessibility:** Addressing disparities in access to advanced technologies and ensuring equitable benefits across socio-economic groups.
- **Human Rights:** Protecting human dignity and preventing exploitation or discrimination based on augmented abilities.

12.1.5 Challenges in Integrating AI and Human Augmentation

Despite the potential benefits, integrating AI and human augmentation faces several challenges:

- **Technological Limitations:** Developing reliable AI algorithms and robust human-machine interfaces for seamless integration.
- **Regulatory Frameworks:** Establishing ethical guidelines and regulatory frameworks to govern AI-augmented technologies and ensure safety and fairness.
- **Social Acceptance:** Overcoming cultural resistance, stigma, and ethical debates surrounding the ethical use of AI and human augmentation.
- **Bias and Discrimination:** Mitigating biases in AI algorithms and addressing concerns about algorithmic fairness and transparency.

12.1.6 Future Directions and Innovations

Future advancements in AI and human augmentation are poised to transform industries and society:

- **Advances in Neural Interfaces:** Enhanced BCIs and neuroprosthetics that integrate seamlessly with neural circuits.
- **Personalized Healthcare:** AI-driven precision medicine and regenerative therapies tailored to individual genetic profiles.
- **Augmented Reality:** AR applications for real-time data visualization, remote collaboration, and immersive experiences.
- **Ethical AI Development:** Promoting responsible AI research and development that prioritizes ethical considerations and societal impact.

12.2 QUANTUM COMPUTING AND HUMAN AUGMENTATION

The convergence of quantum computing and human augmentation represents a transformative frontier at the intersection of advanced computing capabilities and enhancing human capabilities.

Quantum computing offers unprecedented computational power and efficiency, potentially revolutionizing fields from healthcare and artificial intelligence to cryptography and materials science.

The integration of quantum computing with human augmentation represents a paradigm shift in advancing technological capabilities and enhancing human potential across diverse applications.

By leveraging quantum computing's computational power and efficiency, we can unlock new frontiers in healthcare, neuroscience, AI, and beyond, benefiting individuals and society as a whole. Addressing ethical considerations, regulatory challenges, and technological complexities will be essential in realizing the full potential of quantum computing and human augmentation responsibly and ethically. As these technologies evolve, fostering interdisciplinary collaboration, public engagement, and ethical stewardship will pave the way for transformative innovations that empower individuals and elevate humanity's collective capabilities in the quantum era.

12.2.1 Understanding Quantum Computing

Quantum computing harnesses principles of quantum mechanics to process information in ways traditional computers cannot. Unlike classical bits, which represent either 0 or 1, quantum bits (qubits) can exist in multiple states simultaneously due to superposition and entanglement.

This parallel processing capability enables quantum computers to solve complex problems exponentially faster than classical computers, with applications in optimization, cryptography, simulation, and machine learning.

12.2.2 Potential of Quantum Computing in Human Augmentation

Quantum computing holds promise for enhancing human augmentation by:

- **Optimizing Algorithms:** Accelerating the development of AI algorithms for personalized healthcare, cognitive enhancement, and adaptive learning systems.
- **Enhancing Simulation and Modeling:** Facilitating precise simulations of biological systems, drug interactions, and materials properties, advancing medical research and personalized medicine.
- **Improving Data Security:** Strengthening encryption protocols and cybersecurity measures to protect sensitive medical, financial, and personal data.

12.2.3 Applications of Quantum Computing and Human Augmentation

The integration of quantum computing with human augmentation spans various domains:

- **Healthcare:** Quantum computing enhances genomic sequencing, drug discovery, and diagnostic accuracy, revolutionizing personalized medicine and treatment planning.
- **Neural Interfaces:** Quantum algorithms optimize brain-computer interfaces (BCIs) and neuroprosthetics, improving communication and control for individuals with disabilities.
- **Cognitive Enhancement:** Quantum-inspired algorithms support cognitive training, memory enhancement, and decision-making processes.

- **Biomechanical Modeling:** Quantum simulations enable precise modeling of human biomechanics, advancing prosthetic design and rehabilitation strategies.

12.2.4 Ethical Considerations in Quantum Computing and Human Augmentation

Ethical considerations in integrating quantum computing with human augmentation encompass:

- **Privacy and Security:** Protecting sensitive biological and neural data from quantum-enabled cyber threats and breaches.
- **Equity and Access:** Addressing disparities in access to quantum computing resources and ensuring equitable distribution of benefits.
- **Algorithmic Fairness:** Mitigating biases in quantum algorithms and ensuring fairness in AI-driven applications.
- **Human Rights:** Safeguarding autonomy, consent, and dignity in the development and use of quantum-enhanced technologies.

12.2.5 Challenges in Integrating Quantum Computing and Human Augmentation

Despite its potential, integrating quantum computing with human augmentation faces challenges:

- **Technological Complexity:** Developing scalable quantum hardware and robust quantum algorithms for practical applications in human augmentation.
- **Interdisciplinary Collaboration:** Fostering collaboration between quantum physicists, computer scientists, neuroscientists, and healthcare professionals to integrate quantum technologies with human augmentation effectively.
- **Regulatory Frameworks:** Establishing ethical guidelines and regulatory frameworks to govern quantum-enhanced technologies and ensure responsible innovation.
- **Public Perception and Acceptance:** Addressing societal concerns, skepticism, and misconceptions about the implications of quantum computing and human augmentation.

12.2.6 Future Directions and Innovations

Future advancements in quantum computing and human augmentation are poised to:

- **Enhance Precision Medicine:** Quantum-enabled simulations and genomic analysis revolutionize personalized healthcare and treatment strategies.
- **Optimize Brain-Computer Interfaces:** Quantum algorithms improve neural communication, prosthetic control, and assistive technologies.
- **Transform AI and Machine Learning:** Quantum-inspired algorithms enhance machine learning models, natural language processing, and predictive analytics.
- **Expand Scientific Discovery:** Quantum simulations accelerate breakthroughs in neuroscience, materials science, and climate modeling.

12.3 ANTICIPATING THE NEXT WAVE OF INNOVATIONS

Anticipating the next wave of innovations involves exploring emerging technologies, trends, and transformative advancements poised to reshape industries, economies, and societal landscapes.

12.3.1 Drivers of Innovation

Innovation is driven by various factors that catalyze change and spur technological advancements:

- **Technological Advancements:** Breakthroughs in fields like artificial intelligence, quantum computing, biotechnology, and renewable energy expand possibilities for innovation.
- **Market Demand:** Consumer needs and preferences drive innovation in products, services, and user experiences.
- **Regulatory Environment:** Policies and regulations shape innovation incentives, standards, and market dynamics.
- **Global Challenges:** Addressing pressing issues such as climate change, healthcare access, and resource scarcity fuels innovation in sustainable solutions.

12.3.2 Key Technological Trends Shaping the Future

Several technological trends are shaping the landscape of future innovations:

- **Artificial Intelligence and Machine Learning:** Advancements in AI algorithms, deep learning, and natural language processing revolutionize automation, decision-making, and personalized experiences across sectors.
- **Quantum Computing:** Quantum-enabled computing power accelerates complex problem-solving, cryptography, and scientific discoveries with profound implications for industries and scientific research.

- **Biotechnology and Genomics:** Innovations in gene editing, personalized medicine, and bioinformatics drive breakthroughs in healthcare, agriculture, and environmental sustainability.
- **Internet of Things (IoT):** Connected devices, smart cities, and IoT ecosystems enhance efficiency, connectivity, and data-driven insights in urban planning, transportation, and infrastructure.
- **Clean Energy and Sustainability:** Renewable energy technologies, energy storage solutions, and circular economy practices foster sustainable development and mitigate environmental impacts.

12.3.3 Ethical Considerations in Future Innovations

As innovations advance, ethical considerations become increasingly important:

- **Privacy and Data Security:** Safeguarding personal information and sensitive data from breaches, cyber threats, and misuse.
- **Algorithmic Bias:** Mitigating biases in AI algorithms and ensuring fairness, transparency, and accountability in automated decision-making.
- **Human Augmentation:** Balancing ethical implications of enhancing human capabilities through technologies like AI, genetic engineering, and neural interfaces.
- **Environmental Impact:** Promoting sustainable practices, reducing carbon footprints, and addressing ecological consequences of technological advancements.

12.3.4 Challenges Ahead

Anticipating the next wave of innovations is accompanied by challenges that require careful navigation:

- **Technological Risks:** Managing risks associated with emerging technologies, including cybersecurity threats, unintended consequences, and societal disruptions.
- **Regulatory Frameworks:** Establishing adaptive policies and governance structures to foster innovation while ensuring safety, ethical standards, and compliance.
- **Skills Gap:** Addressing the need for workforce training, education, and reskilling to meet demands of evolving technological landscapes.
- **Global Collaboration:** Promoting international cooperation, knowledge sharing, and ethical standards to address global challenges and maximize the benefits of innovations.

12.3.5 Future Implications of Innovations

The future implications of upcoming innovations are profound:

- **Economic Growth:** Stimulating job creation, entrepreneurship, and economic opportunities in emerging industries and digital economies.
- **Healthcare Advancements:** Enhancing disease prevention, diagnostics, treatment outcomes, and healthcare access through personalized medicine and telemedicine.
- **Environmental Sustainability:** Accelerating transitions to clean energy, circular economy models, and sustainable development practices to mitigate climate change impacts.
- **Social Transformation:** Influencing societal behaviors, cultural norms, and governance structures through digital transformation, connectivity, and information accessibility.

12.3.6 Embracing a Future of Innovation

Anticipating the next wave of innovations requires foresight, collaboration, and strategic foresight to navigate opportunities and challenges. By embracing technological advancements responsibly, fostering ethical considerations, and addressing global imperatives, stakeholders can shape a future that is inclusive, sustainable, and equitable. Harnessing the potential of upcoming innovations will empower societies to address complex challenges, drive economic prosperity, and enhance human well-being in an interconnected and rapidly evolving world.

CHAPTER 13

REGULATORY FRAMEWORK AND STANDARDS

13.1 ENSURING SAFETY AND COMPLIANCE

In the realm of technological advancements, human augmentation stands at the forefront of innovation, promising to redefine the capabilities and limitations of the human body.

From prosthetic limbs that provide enhanced mobility to neural implants that augment cognitive functions, the potential benefits are vast. However, alongside these advancements come significant ethical, legal, and safety considerations.

Human augmentation holds immense promise for enhancing human capabilities and improving quality of life. However, realizing this potential requires careful consideration of ethical implications, robust regulatory frameworks, and rigorous safety protocols.

Stakeholders must collaborate to ensure that these technologies are developed responsibly, with a focus on equity, privacy, and public safety. By navigating these challenges thoughtfully and inclusively, we can harness the transformative power of human augmentation while safeguarding the well-being and rights of individuals worldwide.

Human augmentation refers to the use of technology to enhance human capabilities, whether physical, cognitive, or sensory. The technologies involved can range from simple wearable devices to complex implants that integrate with the body's natural systems. Examples include exoskeletons that boost strength and endurance, neural interfaces that enhance brain-to-computer communication, and genetic modifications aimed at improving physiological traits.

As with any transformative technology, human augmentation raises profound ethical questions. Central to these concerns are issues of equity and access—will these technologies exacerbate existing inequalities by favoring those who can afford them? Additionally, ethical debates surround the concept of human identity and autonomy. How much augmentation is too much before it fundamentally alters what it means to be human? Ethicists argue that stringent guidelines must be established to safeguard against potential misuse and unintended consequences.

13.1.1 Regulatory Frameworks and Legal Considerations

The rapid pace of technological development often outpaces regulatory frameworks designed to ensure safety and protect human rights. Establishing clear and comprehensive regulations is paramount to mitigate risks associated with human augmentation. Regulatory bodies worldwide face the challenge of balancing innovation with public safety. Key considerations include device safety standards, data privacy protections, and protocols for clinical trials and post-market surveillance. International cooperation is crucial to harmonize regulations and prevent regulatory arbitrage.

13.1.2 Safety Standards and Testing Protocols

Ensuring the safety and efficacy of augmented technologies requires rigorous testing and adherence to established standards. Medical devices, including those used for human augmentation, must undergo thorough preclinical and clinical trials to assess their performance, reliability, and potential risks. Regulatory agencies such as the FDA in the United States and the EMA in Europe play pivotal roles in evaluating these technologies before they reach the market. Continuous monitoring post-approval is essential to identify and address any unforeseen adverse effects promptly.

13.1.3 Privacy and Data Security

Human augmentation often involves the collection and analysis of sensitive personal data. Neural implants, for instance, may interface directly with the brain's neural networks, raising concerns about data security and privacy breaches. Robust encryption protocols and strict access controls are imperative to safeguard against unauthorized access and exploitation of personal information. Ethical guidelines must also address consent protocols for data collection and use, ensuring individuals retain autonomy over their personal information.

13.1.4 Public Perception and Acceptance

Public perception of human augmentation is shaped by a complex interplay of factors including cultural attitudes, media portrayals, and personal beliefs. Addressing concerns and fostering informed dialogue are essential to garnering widespread acceptance. Transparency in research and development processes, coupled with proactive engagement with stakeholders, can help build trust and mitigate fears of unforeseen consequences. Educational initiatives aimed at raising awareness about the benefits and risks of augmentation technologies are equally critical.

13.1.5 Challenges and Future Directions

Despite the potential benefits of human augmentation, significant challenges lie ahead. Technological risks, ethical dilemmas, and regulatory hurdles necessitate collaborative efforts across disciplines and sectors. Interdisciplinary research initiatives can foster innovation while addressing ethical and safety concerns. Long-term studies are essential to assess the impact of augmentation technologies on individuals and society at large. Moreover, international cooperation is indispensable in developing globally harmonized standards and regulations.

13.2 INTERNATIONAL STANDARD FOR HUMAN AUGMENTATION

Human augmentation, the enhancement of human abilities through technological interventions, has emerged as a transformative field with vast potential and profound ethical implications. As this technology advances rapidly, the need for an international standard governing its development, deployment, and ethical considerations becomes increasingly pressing.

The development of an international standard for human augmentation is essential to navigate the complexities of this transformative technology responsibly. Such a standard would harmonize regulatory frameworks, establish ethical guidelines, and ensure robust safety and privacy protections across borders. While challenges abound, collaborative efforts among stakeholders can pave the way for a future where human augmentation advances human capabilities while upholding universal principles of ethics, safety, and respect for human dignity.

Human augmentation encompasses a wide array of technologies aimed at enhancing human capabilities. These technologies can range from prosthetic limbs and exoskeletons that improve physical abilities to neural implants and genetic modifications that enhance cognitive functions and physiological traits. The goal is to surpass natural human limitations and improve quality of life, but these advancements also raise ethical, legal, and safety concerns that must be addressed through robust standards.

13.2.1 Necessity of an International Standard

In a globally connected world where technological developments transcend borders, establishing an international standard for human augmentation is crucial. Such a standard would harmonize regulatory frameworks across countries, ensuring consistency in safety protocols, ethical guidelines, and legal protections. It would facilitate innovation by providing clear guidelines for researchers, developers, and regulatory agencies, thus promoting public trust and confidence in these transformative technologies.

13.2.2 Key Components of an International Standard

Safety and Efficacy Standards

Central to any international standard for human augmentation are rigorous safety and efficacy standards. These standards would define criteria for testing the safety, reliability, and performance of augmentation technologies before they reach the market. Regulatory agencies would assess compliance with these standards through preclinical and clinical trials, ensuring that devices meet established safety benchmarks.

Ethical Guidelines

Ethical considerations are paramount in human augmentation, given its potential to alter human capabilities and identity. An international standard would outline ethical guidelines to address concerns such as equity in access to augmentation technologies, informed consent for interventions, protection of human rights, and prevention of misuse or exploitation. These guidelines would promote responsible development and deployment of augmentation technologies while safeguarding individual autonomy and dignity.

Privacy and Data Security Protocols

Human augmentation often involves the collection and analysis of sensitive personal data, particularly in technologies like neural implants and biometric enhancements. An international standard would establish robust protocols for data privacy and security, including encryption standards, anonymization practices, and guidelines for consent and data access. These protocols would protect individuals from unauthorized access and ensure that personal information is handled ethically and responsibly.

13.2.3 Challenges in Establishing an International Standard

Diverse Regulatory Landscapes

One of the primary challenges in establishing an international standard for human augmentation is navigating the diverse regulatory landscapes of different countries. Regulations vary significantly in terms of scope, stringency, and enforcement mechanisms, making it challenging to achieve consensus on a unified standard that accommodates global diversity while ensuring consistent safety and ethical protections.

Technological Complexity

The rapid pace of technological advancement in human augmentation presents another challenge. New innovations often outpace regulatory frameworks, requiring agile and adaptive standards that can keep pace with emerging technologies. Balancing innovation with safety and ethical considerations requires ongoing dialogue and collaboration among stakeholders across disciplines and sectors.

Ethical and Cultural Diversity

Ethical perspectives on human augmentation vary widely across cultures and societies. Cultural norms, religious beliefs, and philosophical considerations shape attitudes towards technological interventions in the human body. An international standard must respect and accommodate this diversity while upholding universal principles of human rights, autonomy, and dignity. Achieving consensus on ethical guidelines that are both globally applicable and culturally sensitive poses a significant challenge.

13.2.4 Future Directions and Considerations

Interdisciplinary Collaboration

Addressing the complexities of human augmentation requires interdisciplinary collaboration among scientists, ethicists, policymakers, and stakeholders from diverse backgrounds. Collaborative efforts can facilitate the development of comprehensive standards that integrate scientific knowledge with ethical principles and regulatory expertise. Initiatives such as international conferences, research networks, and joint policy frameworks can promote dialogue and consensus-building on global standards for human augmentation.

Stakeholder Engagement and Public Awareness

Engaging stakeholders and fostering public awareness are essential in shaping an international standard for human augmentation. Stakeholders include patients, healthcare providers, industry representatives, advocacy groups, and regulatory agencies. Transparent communication, inclusive decision-making processes, and educational initiatives can build trust, address concerns, and garner support for ethical and scientifically sound standards.

Continuous Evaluation and Adaptation

Establishing an international standard is not a static process but requires continuous evaluation and adaptation in response to technological advancements, ethical debates, and regulatory developments. Regular review mechanisms, feedback loops, and mechanisms for updating standards based on new evidence and insights are essential to ensure that standards remain relevant, effective, and responsive to evolving challenges and opportunities.

13.3 LEGAL AND REGULATORY CHALLENGES

Human augmentation, the integration of technological enhancements into the human body to improve physical, cognitive, or sensory abilities, presents a myriad of legal and regulatory challenges. As these technologies advance rapidly, governments, international organizations, and regulatory bodies face the daunting task of establishing frameworks that ensure safety, protect human rights, and foster innovation.

Navigating the legal and regulatory challenges of human augmentation requires a balanced approach that promotes innovation while safeguarding public health, privacy, and ethical principles. Establishing robust regulatory frameworks, addressing ethical considerations, and fostering international collaboration are essential steps toward realizing the potential benefits of augmentative technologies responsibly and equitably. By addressing these challenges proactively and inclusively, stakeholders can shape a future where human augmentation enhances quality of life while upholding fundamental rights and values.

Human augmentation encompasses a broad spectrum of technologies and interventions designed to enhance human capabilities beyond natural limits. These technologies include prosthetics, exoskeletons, neural implants, genetic modifications, and wearable devices. While promising significant benefits in terms of mobility, communication, and overall quality of life, human augmentation also raises profound ethical, social, and legal implications that must be carefully navigated.

13.3.1 Regulatory Frameworks: Current Landscape

National Regulations

Regulatory frameworks for human augmentation vary widely across countries. In the United States, for example, the Food and Drug Administration (FDA) oversees the approval of medical devices, including many types of augmentative technologies. Devices undergo rigorous testing to ensure safety and efficacy before they can be marketed and sold to consumers. Similarly, the European Union (EU) regulates medical devices through the Medical Devices Regulation (MDR), which sets standards for safety, performance, and clinical evaluation.

International Standards and Harmonization

While national regulations exist, the lack of harmonization on an international level poses challenges for manufacturers, researchers, and consumers alike. International organizations such as the International Organization for Standardization (ISO) and the World Health Organization (WHO) play crucial roles in developing global standards and guidelines. Harmonizing regulations can streamline market access, facilitate innovation, and ensure consistent safety and ethical standards across borders.

13.3.2 Key Legal and Regulatory Challenges

Safety and Efficacy

Ensuring the safety and efficacy of augmentative technologies is paramount but complex. Unlike traditional medical devices, some augmentation technologies may involve invasive procedures or alter fundamental aspects of human physiology. Regulatory agencies must establish rigorous testing protocols, monitor long-term effects, and update standards as technologies evolve to mitigate risks and protect public health.

Ethical Considerations

Human augmentation raises profound ethical questions regarding autonomy, identity, and social equity.

Ethical guidelines must address issues such as informed consent, privacy protections, fair distribution of benefits, and the potential for enhancement disparities based on socioeconomic status. Balancing individual autonomy with societal implications requires thoughtful consideration and stakeholder engagement to develop ethical frameworks that uphold human rights and dignity.

Privacy and Data Security

Many augmentative technologies involve the collection and processing of sensitive personal data, such as neural activity or genetic information. Protecting privacy and ensuring data security are critical to prevent unauthorized access, misuse, or breaches that could compromise individuals' autonomy and rights.

Robust data protection regulations and encryption standards are essential to build trust and compliance among consumers and stakeholders.

13.3.3 Emerging Technologies and Regulatory Gaps

Rapid Technological Advancements

The pace of technological innovation in human augmentation often outstrips regulatory frameworks designed to oversee these developments. Emerging technologies, such as brain-computer interfaces and gene editing tools, challenge existing definitions of medical devices and therapeutic interventions.

Regulatory agencies must adapt quickly to evaluate novel technologies, address unforeseen risks, and ensure that regulations remain relevant and effective.

Undefined Legal Status

The legal status of human augmentation technologies is often ambiguous, blurring the lines between medical treatments, enhancements, and personal enhancements. Questions arise regarding liability for device malfunctions, compensation for adverse effects, and the scope of regulatory oversight.

Clear definitions and classifications are necessary to establish legal accountability, protect consumer rights, and ensure fair access to safe and effective technologies.

13.3.4 Pathways Forward: Addressing Challenges and Promoting Innovation

Strengthening Regulatory Oversight

To address the complex challenges of human augmentation, regulatory bodies must strengthen oversight through collaboration, transparency, and adaptive governance. Establishing specialized committees or task forces can facilitate interdisciplinary dialogue, assess regulatory gaps, and propose updates to ensure that regulations keep pace with technological advancements.

International Collaboration and Harmonization

International collaboration is essential to harmonize regulatory standards and facilitate global market access for augmentative technologies. Organizations like the ISO and WHO can play pivotal roles in developing consensus-based standards that uphold safety, efficacy, and ethical principles across diverse regulatory environments. Mutual recognition agreements and regulatory convergence initiatives can streamline approval processes while maintaining high standards of consumer protection.

Public Engagement and Education

Public engagement and education are critical to fostering trust, understanding, and acceptance of human augmentation technologies. Governments, healthcare providers, and industry stakeholders should prioritize transparent communication, ethical literacy programs, and inclusive dialogue to address misconceptions, mitigate fears, and empower individuals to make informed decisions about their health and well-being.

CHAPTER 14

COLLABORATIVE LEARNING AND EDUCATION

14.1 AUGMENTED REALITY IN EDUCATION

Augmented Reality (AR), once a futuristic concept, has rapidly become integrated into various aspects of our lives, including education. This technology overlays digital information onto the real world, offering unique opportunities to enhance learning experiences. In education, AR holds the promise of transforming traditional teaching methods into interactive, immersive lessons that engage students in ways previously unimaginable.

Augmented Reality represents a transformative force in education, revolutionizing how students learn, collaborate, and interact with information. By combining digital innovation with pedagogical expertise, educators can create dynamic learning environments that inspire curiosity, foster creativity, and prepare students for success in a rapidly evolving world. As AR continues to evolve, its potential to reshape education for the better remains boundless, offering limitless opportunities for learning and growth.

Augmented Reality enhances the real-world environment by superimposing computer-generated content such as images, videos, or 3D models. Unlike Virtual Reality (VR), which creates entirely virtual environments, AR enriches the existing surroundings with digital elements. This integration allows users to interact with both physical and virtual objects simultaneously, blurring the line between the real and digital worlds.

14.1.1 Applications of AR in Education

- **Interactive Learning:** AR facilitates interactive learning experiences by making abstract concepts tangible. For instance, students can visualize complex biological structures in 3D, rotate them, and explore different layers, enhancing their understanding beyond what traditional textbooks offer.
- **Simulations and Practical Training:** In fields like medicine and engineering, AR simulations provide a safe environment for practical training. Medical students can practice surgeries virtually, and

engineering students can experiment with machinery and prototypes, minimizing risks and costs associated with real-world training.

- **Historical and Cultural Exploration:** AR brings history and culture to life by overlaying historical footage, artifacts, or reenactments onto physical locations. Students can witness historical events unfold in their actual settings, fostering a deeper connection and understanding of the past.
- **Language Learning:** Language acquisition can be significantly enhanced through AR, as it allows learners to interact with virtual translations, practice conversations with simulated speakers, and explore foreign environments in context. This immersive approach accelerates language comprehension and retention.
- **Collaborative Learning:** AR facilitates collaborative learning experiences where students can work together on projects, solve problems in teams, and share insights in real-time. This fosters teamwork and communication skills essential for future careers.

14.1.2 Benefits of Augmented Reality in Education

- **Enhanced Engagement:** AR makes learning more engaging and enjoyable by catering to different learning styles. Visual learners benefit from interactive visualizations, while kinesthetic learners can manipulate virtual objects, improving retention and comprehension.
- **Personalized Learning Paths:** AR adapts to individual learning paces and styles, offering personalized feedback and adaptive learning paths. Students can progress at their own speed, receiving tailored content and challenges based on their performance and understanding.
- **Accessibility:** AR makes education more accessible by breaking down geographical barriers and providing equal learning opportunities to students regardless of their location or physical abilities. Virtual field trips and remote collaborations become feasible, enriching the educational experience.
- **Real-World Application:** By bridging theory with practical application, AR prepares students for real-world challenges and careers. They gain hands-on experience in problem-solving and critical thinking, essential skills in today's dynamic job market.
- **Cost-Effectiveness:** While initially requiring investment in technology, AR can ultimately reduce long-term costs by replacing expensive physical resources such as lab equipment, field trip expenses, and

printed materials. It also minimizes the need for physical space, making education more sustainable.

14.1.3 Challenges and Considerations

- **Technological Barriers:** The effectiveness of AR in education depends on reliable internet connectivity, access to compatible devices, and technical support. Unequal distribution of technology can widen educational disparities, limiting access for underserved communities.
- **Pedagogical Integration:** Integrating AR into curricula requires training educators in new instructional methods and content creation. Teachers must adapt lesson plans to incorporate AR effectively, ensuring it enhances learning outcomes without overshadowing traditional teaching approaches.
- **Privacy and Security:** AR collects data to personalize learning experiences, raising concerns about student privacy and data security. Safeguards must be in place to protect sensitive information and ensure compliance with privacy regulations.
- **Ethical Considerations:** AR content must be culturally sensitive and ethically sound to avoid perpetuating stereotypes or biases. Educators and developers should collaborate to create inclusive and diverse experiences that promote empathy and understanding.

14.1.4 Future Trends and Innovations

The future of AR in education promises continued advancements in content quality, device compatibility, and user interface design. As technology evolves, AR applications will become more seamless, intuitive, and integrated into everyday educational practices.

14.2 HUMAN-MACHINE COLLABORATION IN TRAINING

Human-machine collaboration in training refers to the integration of human expertise with machine capabilities to enhance learning and skill development across various domains. This approach leverages the strengths of both humans and machines, combining human intuition, creativity, and adaptability with the precision, speed, and data processing capabilities of machines. In fields ranging from healthcare and manufacturing to aviation and military operations, human-machine collaboration is reshaping training methodologies and improving outcomes.

Human-machine collaboration in training represents a paradigm shift in how skills and knowledge are acquired and applied across diverse industries. By harnessing the complementary strengths of humans and machines, organizations can innovate training methodologies, improve learning outcomes, and prepare individuals for success in an increasingly technology-driven world.

As technologies evolve and capabilities expand, the potential for human-machine collaboration to revolutionize training and skill development remains limitless, paving the way for a future where human potential is enhanced through intelligent partnerships with machines.

Human-machine collaboration involves synergistic interactions between humans and intelligent machines, such as robots, AI systems, and virtual assistants. Unlike traditional training methods where humans learn independently or from other humans, this approach emphasizes joint efforts where machines support and augment human capabilities to achieve superior performance and efficiency.

14.2.1 Applications of Human-Machine Collaboration in Training

- **Healthcare:** In healthcare, human-machine collaboration enhances medical training and patient care. Simulation technologies combined with AI-driven diagnostics enable medical professionals to practice complex procedures in realistic virtual environments. Robots assist surgeons during surgeries, improving precision and reducing recovery times.
- **Manufacturing and Industry:** Manufacturing and industrial sectors utilize human-machine collaboration to train workers in operating complex machinery safely and efficiently. Virtual reality (VR) and augmented reality (AR) simulations allow workers to practice assembly tasks and troubleshoot equipment issues without the risk of real-world accidents.
- **Aviation and Aerospace:** Pilots and aerospace engineers benefit from human-machine collaboration through flight simulators and predictive maintenance systems. AI algorithms analyze flight data to optimize fuel efficiency and predict mechanical failures, while VR simulations enable pilots to train for various scenarios in realistic cockpit environments.
- **Military Training:** Military training relies heavily on human-machine collaboration for combat simulations, tactical planning, and logistics management. AI-powered simulations replicate battlefield conditions, allowing soldiers to practice decision-making under stress and refine their combat skills in a safe, controlled environment.
- **Education and Skills Development:** In education, human-machine collaboration supports personalized learning experiences and skills development. Intelligent tutoring systems use AI to adapt educational

content based on student performance, providing targeted feedback and learning resources to enhance understanding and retention.

14.2.2 Benefits of Human-Machine Collaboration in Training

- **Enhanced Learning Efficiency:** By combining human expertise with machine precision, training programs become more efficient and effective. Machines can process vast amounts of data, identify patterns, and deliver personalized feedback, accelerating learning curves and improving overall performance.
- **Improved Safety:** Human-machine collaboration reduces risks associated with training in high-stakes environments. Simulation technologies allow trainees to practice hazardous tasks without endangering themselves or others, ensuring they are prepared for real-world challenges.
- **Cost Savings:** Training with machines often reduces costs associated with traditional methods. Virtual simulations eliminate the need for physical resources and minimize downtime for equipment maintenance, making training programs more sustainable and budget-friendly.
- **Scalability:** Machines enable scalable training solutions that can accommodate large numbers of learners simultaneously. Virtual classrooms and online simulations provide flexible learning environments accessible anytime, anywhere, catering to diverse training needs and schedules.
- **Innovation and Adaptability:** Human-machine collaboration fosters innovation by combining human creativity with machine learning capabilities. Training programs evolve with technological advancements, incorporating new tools and techniques to stay ahead of industry trends and challenges.

14.2.3 Challenges and Considerations

- **Technological Integration:** Effective human-machine collaboration requires seamless integration of technologies into existing training frameworks. Compatibility issues, cybersecurity concerns, and the need for specialized training can hinder adoption and implementation.
- **Ethical and Social Implications:** The use of AI and robotics in training raises ethical questions regarding job displacement, privacy, and autonomy. Balancing technological advancements with ethical considerations ensures that human values and rights are respected in collaborative environments.

- **Skills Gap and Training Needs:** Adopting human-machine collaboration requires upskilling and reskilling of workforce to effectively utilize new technologies. Addressing skills gaps and training needs ensures that individuals and organizations can fully leverage the benefits of collaborative training approaches.
- **Regulatory Compliance:** Industries must adhere to regulatory standards and safety protocols when implementing human-machine collaboration in training. Compliance with industry regulations ensures the legality and ethical use of technology in educational and professional settings.

14.2.4 Future Trends and Innovations

The future of human-machine collaboration in training holds promise for continued advancements in AI, robotics, and virtual technologies. Enhanced machine learning algorithms, interactive simulations, and adaptive learning platforms will further personalize training experiences and optimize performance outcomes.

14.3 THE ROLE OF AUGMENTATION IN LIFELONG LEARNING

Lifelong learning, once viewed as a personal commitment to continuous education beyond formal schooling, is increasingly influenced by augmentation technologies. Augmentation in lifelong learning refers to the integration of digital tools, AI-driven systems, and augmented reality (AR) to enhance and extend learning capabilities throughout one's life.

Augmentation technologies are transforming lifelong learning by expanding educational opportunities, enhancing learning experiences, and empowering individuals to thrive in a knowledge-driven economy. By embracing these technologies, learners can navigate career transitions, acquire new competencies, and pursue lifelong personal and professional growth. As augmentation technologies evolve and integrate into educational ecosystems, their potential to revolutionize lifelong learning remains profound, paving the way for a future where learning knows no boundaries.

14.3.1 Understanding Augmentation in Lifelong Learning

Augmentation in lifelong learning encompasses technologies that supplement human abilities and extend cognitive functions. These tools include AI-powered adaptive learning platforms, virtual reality (VR) simulations, augmented reality (AR) applications, and personalized learning algorithms. By leveraging these technologies, individuals can personalize their learning experiences, access real-time information, and collaborate globally, transcending traditional educational boundaries.

14.3.2 Applications of Augmentation in Lifelong Learning

- **Personalized Learning Paths:** Augmentation technologies analyze individual learning behaviors and preferences to tailor educational content and pace. AI algorithms recommend personalized courses, modules, and resources that align with learners' interests and career goals, fostering self-directed learning and continuous improvement.
- **Skill Acquisition and Development:** In today's rapidly evolving job market, augmentation technologies facilitate skill acquisition and development. VR simulations allow learners to practice complex tasks in realistic environments, while AR overlays digital information onto physical surroundings, enhancing hands-on training and practical knowledge application.
- **Professional Development:** Professionals benefit from augmentation technologies for continuous professional development (CPD). Online platforms offer micro-learning modules, webinars, and virtual workshops that enable professionals to stay updated on industry trends, expand their skill sets, and earn certifications remotely.
- **Global Collaboration:** Augmentation technologies promote global collaboration among learners, educators, and experts. Virtual classrooms, online forums, and collaborative tools facilitate knowledge sharing, cultural exchange, and interdisciplinary research, fostering a diverse and inclusive learning community.
- **Enhanced Accessibility:** Augmentation technologies democratize access to education by overcoming geographical barriers and physical limitations. Mobile apps, cloud-based platforms, and digital libraries provide on-demand learning resources, ensuring equitable educational opportunities for learners worldwide.

14.3.3 Benefits of Augmentation in Lifelong Learning

Flexibility and Convenience

Augmentation technologies offer flexible learning options that accommodate diverse schedules and learning preferences.

Learners can access educational content anytime, anywhere, using devices such as smartphones, tablets, and wearable technologies, enhancing convenience and accessibility.

Personalized Learning Experiences

Augmentation technologies adapt educational content to individual learning styles and proficiency levels, promoting personalized learning experiences. AI-driven analytics track learner progress, identify areas for improvement, and provide targeted feedback, optimizing learning outcomes.

Continuous Adaptation to Industry Trends

Lifelong learners equipped with augmentation technologies can continuously adapt to industry trends and technological advancements.

Real-time updates, industry insights, and predictive analytics empower learners to anticipate future skills demands and proactively acquire relevant competencies.

Cost-Effective Education

Augmentation technologies reduce costs associated with traditional learning methods, such as travel expenses, physical textbooks, and classroom infrastructure. Online courses, open educational resources (OER), and digital subscriptions offer affordable learning alternatives that support lifelong learning initiatives.

Empowerment and Self-Efficacy

Augmentation technologies empower individuals to take ownership of their learning journeys and develop self-efficacy.

Interactive simulations, gamified learning experiences, and virtual mentors encourage active participation, experimentation, and problem-solving skills development.

14.3.4 Challenges and Considerations

Digital Literacy and Skills Acquisition

Effective use of augmentation technologies requires proficiency in digital literacy and skills acquisition. Addressing digital divides and providing training programs on technology use ensure equitable access and participation in lifelong learning opportunities.

Privacy and Data Security

Augmentation technologies collect and analyze user data to personalize learning experiences and improve system performance.

Safeguarding privacy, securing sensitive information, and complying with data protection regulations are essential to maintaining trust and ethical standards in lifelong learning environments.

Technological Infrastructure and Accessibility

Unequal access to technological infrastructure, such as high-speed internet connectivity and compatible devices, limits the reach and effectiveness of augmentation technologies in underserved communities. Bridging digital divides and improving technological accessibility are critical for promoting inclusive lifelong learning initiatives.

Integration with Traditional Learning Approaches

Augmentation technologies must complement rather than replace traditional learning approaches. Balancing virtual and physical learning experiences, integrating new technologies into existing curricula, and training educators on technology integration ensure cohesive and effective lifelong learning environments.

14.3.5 Future Trends and Innovations

The future of augmentation in lifelong learning holds promise for advancements in AI, machine learning, and augmented reality technologies. Predictive analytics, adaptive learning algorithms, and immersive simulations will continue to enhance personalized learning experiences, facilitate global collaboration, and promote lifelong learning as a cornerstone of individual and societal development.

CHAPTER 15
REFLECTIONS AND FUTURE PROSPECTS

15.1 LESSONS FROM EARLY ADOPTERS

Human augmentation and human-machine collaboration represent cutting-edge fields where technology enhances human capabilities in various domains. Early adopters of these technologies provide valuable insights into their benefits, challenges, and ethical implications. From prosthetics and exoskeletons to AI-driven cognitive enhancements, early adopters are paving the way for a future where humans and machines collaborate seamlessly to achieve extraordinary feats.

Early adopters of human augmentation and human-machine collaboration technologies provide valuable lessons on innovation, integration, and ethical considerations. By enhancing human capabilities, improving quality of life, and advancing societal acceptance, early adopters pave the way for a future where technology empowers individuals, organizations, and communities to achieve extraordinary feats and shape a more inclusive and prosperous society. As technology continues to evolve, the lessons learned from early adopters will inform ethical guidelines, regulatory policies, and collaborative efforts that promote responsible innovation and harness technology's transformative potential for the benefit of humanity.

Human augmentation involves enhancing human capabilities through technological interventions, such as prosthetics, implants, and cognitive enhancements. These technologies aim to improve physical, cognitive, or sensory functions, restoring abilities lost due to injury or enhancing natural capabilities beyond biological limitations. Human-machine collaboration, on the other hand, integrates human expertise with machine intelligence to achieve superior performance and efficiency in various tasks and industries.

15.1.1 Early Adopters of Human Augmentation Technologies

- **Prosthetics and Bionics:** Early adopters of prosthetics and bionics include individuals with limb loss or limb differences. Advanced prosthetic limbs equipped with sensors and actuators mimic natural movements, restoring mobility and functionality. Bionic technologies, such as mind-controlled prosthetics, enable intuitive control and seamless integration with the user's nervous system, improving quality of life and independence.

- **Exoskeletons:** Exoskeletons are wearable robotic devices that augment human strength and endurance. Early adopters include workers in industries like construction and manufacturing, where exoskeletons reduce physical strain, prevent injuries, and enhance productivity. Military personnel also use exoskeletons for carrying heavy loads and navigating challenging terrains, demonstrating their potential for enhancing human performance in demanding environments.
- **Neurotechnologies:** Neurotechnologies focus on enhancing cognitive functions through brain-computer interfaces (BCIs) and neuroprosthetics. Early adopters include individuals with neurological conditions such as paralysis or ALS (Amyotrophic Lateral Sclerosis). BCIs enable direct communication between the brain and external devices, empowering users to control computers, robotic arms, or assistive technologies through neural signals alone.

15.1.2 Early Adopters of Human-Machine Collaboration Technologies

- **AI and Machine Learning:** Early adopters of AI and machine learning technologies leverage data-driven insights to enhance decision-making, automate routine tasks, and improve operational efficiency. Industries such as healthcare, finance, and logistics use AI algorithms for diagnosing diseases, predicting market trends, and optimizing supply chain management, demonstrating AI's transformative impact on human-machine collaboration.
- **Robotics and Automation:** Robotics and automation technologies are early adopters in manufacturing, agriculture, and logistics sectors. Industrial robots perform repetitive tasks with precision and consistency, augmenting human labor and increasing production efficiency. Collaborative robots (cobots) work alongside human workers, enhancing safety and productivity in shared workspaces.
- **Virtual and Augmented Reality:** Virtual reality (VR) and augmented reality (AR) technologies are early adopters in training, education, and entertainment industries. VR simulations immerse users in realistic environments for training purposes, such as medical procedures or military simulations. AR overlays digital information onto the physical world, enhancing real-time decision-making and interactive learning experiences.

15.1.3 Lessons Learned from Early Adopters

- **Enhanced Human Capabilities:** Early adopters demonstrate that human augmentation and human-machine collaboration technologies enhance human capabilities beyond traditional limits. Prosthetics restore mobility and independence, exoskeletons augment physical strength, and AI improves decision-making and task efficiency, illustrating the potential for technology to empower individuals and organizations.
- **Improved Quality of Life:** For individuals with disabilities or injuries, human augmentation technologies improve quality of life by restoring lost functions and promoting independence. Prosthetics and neurotechnologies enable users to perform daily activities and pursue personal and professional goals, highlighting the transformative impact of technology on human well-being.
- **Ethical Considerations:** Early adopters emphasize the importance of addressing ethical considerations in the development and deployment of human augmentation and human-machine collaboration technologies. Issues such as privacy, data security, informed consent, and equitable access to technology must be prioritized to ensure responsible innovation and mitigate potential risks.
- **Integration and Adaptation:** Successful adoption of human augmentation and human-machine collaboration technologies requires seamless integration into existing workflows, training programs, and regulatory frameworks. Early adopters emphasize the importance of interdisciplinary collaboration, user feedback, and continuous adaptation to optimize technology deployment and maximize benefits.

15.1.4 Challenges and Considerations

- **Technological Advancements:** Rapid technological advancements in human augmentation and human-machine collaboration pose challenges related to safety, reliability, and interoperability. Early adopters advocate for robust testing protocols, regulatory oversight, and standards compliance to ensure the ethical and effective use of emerging technologies.
- **Societal Acceptance:** Societal acceptance of human augmentation and human-machine collaboration technologies varies based on cultural norms, ethical beliefs, and perceptions of risk. Early adopters emphasize the importance of education, public engagement, and transparent communication to foster understanding and acceptance of technology's potential benefits and risks.

- **Equitable Access:** Ensuring equitable access to human augmentation and human-machine collaboration technologies is essential for addressing socioeconomic disparities and promoting inclusive innovation. Early adopters advocate for affordability, accessibility, and policy initiatives that prioritize marginalized communities and underserved populations.

15.1.5 Future Directions and Innovations

The future of human augmentation and human-machine collaboration promises continued advancements in AI, robotics, neurotechnologies, and wearable devices. Early adopters envision personalized medicine, enhanced human-machine interfaces, and ethical frameworks that guide responsible innovation and maximize societal benefits.

15.2 ANTICIPATING SOCIAL AND ECONOMIC TRANSFORMATIONS

Anticipating social and economic transformations involves understanding the dynamic interplay of factors such as technological advancements, demographic shifts, geopolitical changes, and cultural evolutions. By analyzing current trends and emerging developments, societies can prepare for future challenges and opportunities, ensuring resilience and sustainable growth in a rapidly changing world.

Anticipating social and economic transformations requires foresight, collaboration, and adaptive strategies to navigate complex global dynamics and emerging challenges. By embracing technological advancements, addressing demographic shifts, and fostering inclusive development, societies can build resilient economies, promote sustainable practices, and create equitable opportunities for future generations. As stakeholders across sectors collaborate to shape a more prosperous and interconnected world, proactive anticipation of transformations ensures readiness to seize opportunities and overcome challenges in a rapidly evolving global landscape.

15.2.1 Technological Advancements and Their Impact

- **Artificial Intelligence and Automation:** Artificial Intelligence (AI) and automation are reshaping industries by optimizing processes, increasing productivity, and transforming business models. As AI algorithms become more sophisticated, they automate routine tasks, enhance decision-making capabilities, and create new job opportunities in emerging fields such as data science and machine learning.
- **Internet of Things (IoT) and Connectivity:** The Internet of Things (IoT) connects devices, vehicles, and infrastructure, enabling real-time data exchange and remote control. IoT technologies improve efficiency

in sectors like healthcare, agriculture, and smart cities, enhancing resource management, environmental sustainability, and urban planning.

- **Blockchain and Decentralization:** Blockchain technology facilitates secure, transparent transactions without intermediaries, revolutionizing finance, supply chain management, and digital identity verification. Decentralized applications (DApps) promote peer-to-peer interactions, financial inclusion, and transparent governance models, disrupting traditional industries and fostering innovation.

15.2.2 Demographic Shifts and Societal Changes

- **Aging Population and Healthcare:** An aging population poses challenges and opportunities for healthcare systems, retirement planning, and eldercare services. Innovations in medical technology, personalized medicine, and remote healthcare delivery address aging-related health issues while promoting healthy aging and improving quality of life for seniors.
- **Urbanization and Infrastructure:** Rapid urbanization necessitates sustainable infrastructure development, efficient transportation systems, and resilient urban planning strategies. Smart cities leverage IoT, AI, and renewable energy solutions to optimize resource allocation, reduce carbon footprints, and enhance urban livability for growing populations.
- **Globalization and Cultural Exchange:** Globalization facilitates cultural exchange, economic interdependence, and international cooperation. Cross-cultural collaborations in arts, education, and technology promote diversity, innovation, and mutual understanding, shaping global identities and fostering interconnected communities.

15.2.3 Geopolitical Dynamics and Economic Trends

- **Trade Policies and Global Markets:** Geopolitical tensions influence trade policies, economic stability, and global markets. Shifts in international relations, tariff regulations, and supply chain disruptions impact industries, consumer behavior, and investment strategies, necessitating adaptive strategies and risk management frameworks.
- **Environmental Sustainability and Green Technologies:** Environmental sustainability initiatives promote renewable energy adoption, resource conservation, and climate resilience. Green technologies, such as clean energy solutions and eco-friendly practices,

mitigate environmental impacts, foster sustainable development, and address global challenges like climate change and biodiversity loss.

- **Workforce Trends and Skills Development:** Workforce trends, including remote work, gig economy growth, and automation, redefine labor markets and skills requirements. Lifelong learning, upskilling, and reskilling programs equip individuals with in-demand competencies, promote economic mobility, and support inclusive growth amid technological advancements and economic transformations.

15.2.4 Anticipating Future Challenges and Opportunities

- **Emerging Risks and Resilience Strategies:** Anticipating future challenges involves identifying emerging risks, such as cybersecurity threats, economic volatility, and public health crises. Resilience strategies encompass scenario planning, adaptive governance, and robust infrastructure investments to mitigate risks and ensure continuity in uncertain environments.
- **Innovation and Technological Integration:** Promoting innovation and technological integration fosters economic diversification, industry competitiveness, and job creation. Investments in research and development, digital infrastructure, and entrepreneurship ecosystems drive technological advancements, spur innovation clusters, and position economies for sustainable growth.
- **Social Equity and Inclusive Development:** Promoting social equity involves addressing disparities in access to education, healthcare, and economic opportunities. Inclusive development initiatives prioritize marginalized communities, promote gender equality, and enhance social cohesion through equitable policies, community engagement, and empowerment programs.

15.2.5 Ethical Considerations and Policy Frameworks

- **Ethical Use of Technology:** Ethical considerations in technological advancements include privacy protection, data governance, and responsible AI deployment. Regulatory frameworks, ethical guidelines, and stakeholder engagement ensure transparent practices, safeguard human rights, and promote trust in digital ecosystems.
- **Policy Innovation and Adaptive Governance:** Policy innovation and adaptive governance frameworks facilitate agile responses to evolving social, economic, and environmental challenges. Collaborative policymaking, evidence-based decision-making, and public-private

partnerships enable effective regulation, promote innovation ecosystems, and drive sustainable development agendas.

- **Global Collaboration and Diplomacy:** Global collaboration and diplomacy foster international cooperation, conflict resolution, and collective action on global challenges. Multilateral agreements, diplomatic dialogues, and cross-border partnerships promote peace, security, and sustainable development goals, ensuring shared prosperity in an interconnected world.

15.3 SHAPING A HUMAN-CENTRIC FUTURE

Shaping a human-centric future involves prioritizing human well-being, values, and aspirations in the design and implementation of technological advancements, social policies, and economic systems. By placing humans at the center of decision-making processes, societies can foster inclusive growth, promote sustainable development, and address global challenges while ensuring dignity, equity, and prosperity for all individuals.

Shaping a human-centric future requires collective efforts, ethical leadership, and innovative strategies to prioritize human well-being, equity, and sustainability in a rapidly evolving world. By harnessing technological advancements, promoting inclusive socioeconomic policies, and fostering global collaboration, societies can build resilient communities, empower individuals, and create opportunities for prosperity and dignity for all. As stakeholders across sectors collaborate to shape a more inclusive and sustainable future, proactive measures and human-centered approaches ensure resilience, responsiveness, and readiness to address emerging challenges and seize opportunities in a dynamic global landscape.

15.3.1 Human-Centric Technology and Innovation

- **Ethical Design Principles:** Human-centric technology emphasizes ethical design principles that prioritize user safety, privacy, and autonomy. Technologies such as artificial intelligence (AI), blockchain, and biotechnology are developed with human values in mind, ensuring transparency, fairness, and accountability in their application and deployment.
- **User-Centered Design:** User-centered design methodologies involve engaging stakeholders in the development process to understand their needs, preferences, and challenges. By incorporating user feedback and iterative testing, technologies and products are tailored to enhance user experience, usability, and accessibility for diverse populations.
- **Technological Empowerment:** Technological empowerment aims

to democratize access to information, resources, and opportunities through digital literacy, connectivity, and inclusive design. Empowering individuals and communities with technology promotes economic participation, social inclusion, and civic engagement, driving equitable development and collective prosperity.

15.3.2 Socioeconomic Policies for Inclusive Growth

- **Equitable Access to Education and Skills Development:** Equitable access to education and lifelong learning opportunities ensures individuals acquire essential knowledge, skills, and competencies for personal growth and career advancement. Investing in educational infrastructure, teacher training, and digital literacy programs prepares individuals for future employment and leadership roles in a rapidly evolving global economy.
- **Social Protection and Healthcare:** Social protection systems provide safety nets, healthcare services, and financial assistance to vulnerable populations, promoting well-being, resilience, and social cohesion. Universal healthcare coverage, disability rights protections, and eldercare support contribute to equitable health outcomes and inclusive societies.
- **Fair Labor Practices and Economic Opportunities:** Fair labor practices, including decent wages, workplace safety standards, and labor rights protections, ensure dignity, fairness, and economic security for workers. Promoting entrepreneurship, innovation ecosystems, and sustainable business practices fosters job creation, economic mobility, and wealth distribution in diverse communities.

15.3.3 Environmental Sustainability and Responsible Stewardship

- **Climate Action and Green Technologies:** Climate action initiatives promote environmental sustainability through renewable energy adoption, carbon emissions reduction, and conservation efforts. Investing in green technologies, sustainable infrastructure, and eco-friendly practices mitigates climate change impacts, preserves natural resources, and supports global conservation efforts.
- **Circular Economy and Resource Efficiency:** The circular economy model emphasizes resource efficiency, waste reduction, and product lifecycle sustainability. Implementing circularity principles in manufacturing, consumption patterns, and supply chain management minimizes environmental footprint, fosters resource conservation, and promotes economic resilience in a resource-constrained world.

- **Biodiversity Conservation and Ecosystem Preservation:** Biodiversity conservation efforts protect ecosystems, wildlife habitats, and natural landscapes critical to ecological balance and human well-being. Preserving biodiversity through conservation policies, wildlife sanctuaries, and sustainable land management practices safeguards ecosystem services, promotes ecological resilience, and supports biodiversity-rich regions.

15.3.4 Ethical Leadership and Global Collaboration

- **Ethical Governance and Policy Innovation:** Ethical leadership entails prioritizing integrity, transparency, and accountability in governance and decision-making processes. Implementing ethical frameworks, regulatory standards, and compliance measures ensures responsible stewardship of resources, protects human rights, and fosters public trust in institutions.
- **International Cooperation and Diplomacy:** International cooperation and diplomacy promote global solidarity, peacebuilding, and collective action on shared challenges. Multilateral agreements, diplomatic dialogues, and cross-border partnerships address global issues such as climate change, poverty eradication, and humanitarian crises, advancing sustainable development goals and fostering inclusive global governance.
- **Cultural Diversity and Inclusivity:** Cultural diversity and inclusivity celebrate human heritage, traditions, and identities while promoting mutual respect, tolerance, and dialogue among diverse communities. Embracing cultural diversity through cultural exchanges, intercultural dialogue, and cultural heritage preservation enriches societies, promotes social cohesion, and fosters a sense of belonging in a globalized world.

GLOSSARY OF TERMS

Augmented Reality (AR): Technology that overlays digital information (such as images, text, or animations) onto the real-world environment, enhancing perception and interaction with the physical world.

Bionics: The application of biological principles to the design and creation of artificial limbs, organs, or other body parts, often using electronic or mechanical components.

Brain-Computer Interface (BCI): A direct communication pathway between the brain and an external device, enabling control or interaction based on neural activity.

Cyborg: A cybernetic organism, typically a human with enhanced abilities due to integrated technology, such as implants or prosthetics.

Exoskeleton: A wearable robotic device that supports and enhances human physical capabilities, such as strength, endurance, and mobility.

Human Augmentation: The enhancement of human capabilities through technological means, including physical, cognitive, and sensory enhancements.

Human-Machine Collaboration: The synergy between human expertise and machine capabilities to achieve tasks more efficiently and effectively than either could alone.

Neuroprosthetics: Artificial devices or implants that replace or enhance the functionality of the nervous system, such as prosthetic limbs controlled by neural signals.

Prosthetics: Artificial devices designed to replace or enhance missing or impaired body parts, improving mobility, functionality, or appearance.

Virtual Reality (VR): A computer-generated environment that simulates physical presence in a real or imagined world, often experienced through immersive headsets or displays.

Wearable Technology: Devices worn on the body that incorporate advanced electronics and sensors to monitor health, enhance performance, or provide augmented capabilities.

Human-In-The-Loop (HITL): A concept where humans are actively involved in decision-making processes alongside AI or automated systems, ensuring human oversight and intervention when necessary.

Cognitive Enhancement: Techniques or technologies aimed at improving cognitive functions such as memory, attention, and problem-solving abilities through external means, often involving AI or neurotechnologies.

Robotics: The design, creation, and use of robots to perform tasks traditionally done by humans, ranging from industrial automation to assistive robotics in healthcare.

Teleoperation: The remote control or operation of a device or robot from a distance, often used in hazardous environments or space exploration.

AI Ethics: Principles and guidelines governing the development, deployment, and use of artificial intelligence to ensure ethical behavior, transparency, and accountability.

Machine Learning: A subset of AI that enables machines to learn and improve from experience without explicit programming, using algorithms and data to make predictions or decisions.

Assistive Technology: Devices, tools, or software designed to assist individuals with disabilities, enabling them to perform tasks that might otherwise be difficult or impossible.

Biohacking: The practice of modifying or enhancing biological systems or functions through DIY methods, often involving technology or self-experimentation.

Ethics of Enhancement: Ethical considerations surrounding the use of technologies to enhance human capabilities, including fairness, consent, equality, and societal implications.

RESOURCES FOR FURTHER EXPLORATION

A Cyborg Manifesto: Science, Technology, and Socialist-Feminism in the Late Twentieth Century by Donna Haraway, University of Minnesota Press, 1991.

Artificial Intelligence: A Modern Approach by Stuart Russell and Peter Norvig, Pearson, 2020.

Artificial Intelligence: Foundations of Computational Agents by David L. Poole and Alan K. Mackworth, Cambridge University Press, 2017.

Augmented Human: How Technology Is Shaping the New Reality by Helen Papagiannis, O'Reilly Media, 2017.

Augmented Reality: Principles and Practice by Dieter Schmalstieg and Tobias Hollerer, Addison-Wesley Professional, 2016.

Augmented Reality: Where We Will All Live by Tom Furness, CRC Press, 2017.

Bioengineering: Fundamentals by Ann Saterbak et al., Cambridge University Press, 2017.

Biomechanics and Robotics edited by Gordon A. Rix and Saeid Nahavandi, Springer, 2013.

Biomechatronics: Medicine and Engineering edited by David L. Simpson and Gerard A. Khalil, Woodhead Publishing, 2012.

Biomedical Engineering: Bridging Medicine and Technology by W. Mark Saltzman, Cambridge University Press, 2015.

Cybernetics: Or Control and Communication in the Animal and the Machine by Norbert Wiener, MIT Press, 1961.

Digital Human Modeling: Trends in Human Algorithms edited by Vincent G. Duffy, CRC Press, 2013.

Engineering and Scientific Computing with Scilab by Claude Gomez, Springer, 2006.

Engineering Design and Rapid Prototyping by Ali K. Kamrani and Emad Abouel Nasr, Springer, 2008.

Engineering Principles in Biotechnology edited by Wei-Shou Hu and Michael L. Shuler, Springer, 1993.

Exoskeletons in Rehabilitation Robotics: Tremor Suppression by José L. Pons, Springer, 2018.

Handbook of Virtual Environments: Design, Implementation, and Applications edited by Kay M. Stanney, Lawrence Erlbaum Associates, 2002.

Human Enhancement edited by Julian Savulescu and Nick Bostrom, Oxford University Press, 2009.

Human Factors in Augmented Reality Environments edited by Emily S. Patterson, Springer, 2018.

Human-Computer Interaction and Operators' Performance: Optimizing Work Design with Activity Theory by Brian Gaines, Lawrence Erlbaum Associates, 1992.

Human-Computer Interaction Handbook: Fundamentals, Evolving Technologies, and Emerging Applications edited by Julie A. Jacko and Andrew Sears, CRC Press, 2012.

Human-Computer Interaction: Overview on State of the Art edited by Marco A. Winckler and Peter Forbrig, Springer, 2008.

Human-Machine Reconfigurations: Plans and Situated Actions by Lucy Suchman, Cambridge University Press, 2006.

Human-Robot Interaction edited by Takayuki Kanda et al., Springer, 2008.

Human-Robot Interaction in Social Robotics edited by Takayuki Kanda, Springer, 2012.

Human-Robot Interaction: A Brief History of the Future by William R. Powers, Future Technology Press, 2020.

Human-Robot Interaction: A Survey by Christoph Bartneck et al., Now Publishers Inc, 2009.

Introduction to Artificial Intelligence by Wolfgang Ertel, Springer, 2017.

Introduction to Bioinformatics by Arthur M. Lesk, Oxford University Press, 2017.

Machine Learning for Healthcare: Case Studies and Algorithms for Working with Data by Casey Bennett and Kris Bennett, O'Reilly Media, 2019.

Machine Learning: A Probabilistic Perspective by Kevin P. Murphy, MIT Press, 2012.

Neural Engineering by Bin He, Springer, 2005.

Neuroethics: An Introduction with Readings edited by Martha J. Farah, MIT Press, 2010.

Neuroprosthetics: Theory and Practice edited by Kenneth W. Horch and Gurpreet S. Dhillon, World Scientific Publishing, 2004.

Our Robots, Ourselves: Robotics and the Myths of Autonomy by David A. Mindell, Viking, 2015.

Robot Ethics: The Ethical and Social Implications of Robotics edited by Patrick Lin, Keith Abney, and George A. Bekey, MIT Press, 2012.

Robot Learning edited by Martin Riedmiller and Josef Pauli, Springer, 2000.

Robotics: Modelling, Planning and Control by Bruno Siciliano and Lorenzo Sciavicco, Springer, 2009.

Superintelligence: Paths, Dangers, Strategies by Nick Bostrom, Oxford University Press, 2014.

The Age of Em: Work, Love, and Life when Robots Rule the Earth by Robin Hanson, Oxford University Press, 2016.

The Cyborg Handbook edited by Chris Hables Gray, Routledge, 1995.

The Fourth Industrial Revolution by Klaus Schwab, Crown Business, 2017.

The Human Use of Human Beings: Cybernetics and Society by Norbert Wiener, Da Capo Press, 1988.

The Oxford Handbook of Human-Machine Interaction edited by Harold Bekkering and Paul Haselager, Oxford University Press, 2018.

The Second Machine Age: Work, Progress, and Prosperity in a Time of Brilliant Technologies by Erik Brynjolfsson and Andrew McAfee, W. W. Norton & Company, 2016.

The Singularity Is Near: When Humans Transcend Biology by Ray Kurzweil, Viking Penguin, 2005.

The Society of Mind by Marvin Minsky, Simon & Schuster, 1986.

The Transhumanist Reader: Classical and Contemporary Essays on the Science, Technology, and Philosophy of the Human Future edited by Max More and Natasha Vita-More, Wiley-Blackwell, 2013.

Virtual Reality: Concepts and Technologies by William R. Sherman and Alan B. Craig, CRC Press, 2003.

Wearable Robots: Biomechatronic Exoskeletons by José L. Pons, Wiley-Blackwell, 2008.

INDEX

www.ingramcontent.com/pod-product-compliance
Lightning Source LLC
LaVergne TN
LVHW062116200926
848484LV00007B/1689
9781787154032